INTEGRATED WASTE MANAGEMENT APPROACHES FOR FOOD AND AGRICULTURAL BYPRODUCTS

INTEGRATED WASTE MANAGEMENT APPROACHES FOR FOOD AND AGRICULTURAL BYPRODUCTS

Prof S. K. Singh

Delve
PUBLISHING

www.delvepublishing.com

Integrated Waste Management Approaches for Food and Agricultural by Products

Prof S. K. Singh

Delve Publishing
224 Shoreacres Road
Burlington, ON L7L 2H2
Canada
www.delvepublishing.com

Email: orders@arclereducation.com

ISBN: 978-1-77469-922-5 (Hardcover)

ABOUT THE AUTHOR

Prof. S. K. Singh is the Vice Chancellor of Rajasthan Technical University Kota and a former Professor & Dean at Delhi Technological University. He has a distinguished academic background, with a Ph.D. from BITS Pilani, M. Tech. from IIT BHU Varanasi, and B.E. from Gorakhpur University. He has over 35 years of experience in teaching, research, administration, and consultancy. Prof. Singh has published over 266 research papers, authored 10 books, and received 3 patents in the field of Environmental Engineering. He has guided 12 Ph.D. students and around 80 M.Tech theses. Prof. Singh has held various positions of responsibility, including Director of WAPCOS Limited and Head of Environmental Engineering Department at Delhi Technological University. He has received numerous awards and honors for his contributions to environmental conservation and protection. Prof. Singh has served on expert panels and committees for tasks such as evaluating projects, determining polluting industries, and selecting faculty members. He is also a member of various boards and governing bodies in prestigious institutions as well as a fellow of several professional bodies.

TABLE OF CONTENTS

LIST OF FIGURES

LIST OF ABBREVIATIONS

FAO	Food and Agriculture Organization
SDG	Sustainable Development Goals
UNFCCC	United Nations Framework Convention on Climate Change
GHG	Greenhouse gases
AFW	Agriculture food waste
EPA	Environmental Protection Agency
CNG	Compressed Natural Gas
CCU	Carbon capture and utilization
FW	Food waste
FSC	Food supply chains
AD	Anaerobic digestion
VFA	Volatile fatty acids
FWH	Food waste hydrolysate
SFA	Saturated fatty acids
GT	Gallo-tannins
GRAS	Generally recognized as safe
GR	Ghee residue
SSF	Solid-state fermentation
SmF	Submerged Fermentation
RHA	Rice husk ash
CCU	Carbon capture and utilization
MSW	Municipal solid waste

WFD	Waste Framework Directive
WRAP	Water Reform Action Plan
FFV	Fresh Fruit and Vegetables
UNECE	United Nations Economic Commission for Europe
MRL	Maximum Residue Levels
FEBA	Food Banks Federation
CAA	Clean Air Act
VOC	Volatile organic compounds
BOD	Biochemical oxygen demand
PCB	Polychlorinated biphenyls
PCDD	Polychlorinated dibenzodioxins
PCDF	Polychlorinated dibenzofurans
DDT	Dichlorodiphenyltrichloroethane
COD	Chemical Oxygen Demand
HRT	Hydraulic retention time
TCDD	2,3,7,8-tetrachlorodibenzo-p-dioxin
UASB	Upflow anaerobic sludge blanket reactor
PTA	Purified terephthalic acid
CHP	Combined heat and power
EIA	Environment Impact Assessment
EIS	Environmental impact statement
EMS	Environmental Management System
EP	Economic performance
LCA	Life cycle assessment
OECD	Organisation for Economic Co-operation and Development
OMW	Olive mill wastewater
HPLC	High-performance liquid chromatography (HPLC)
DCFDA	Dichlorodihydrofluorescein diacetate

BHT	Butylated hydroxytoluene
LDL	Low-density lipoprotein
RGM	Red grape marc
WP	Wine pomace
SkWPP	Skins wine pomace product
DPPH	1,1-diphenyl-2-picrylhydrazyl
MDA	Malondialdehyde

PREFACE

The food industry has improved its waste management practices due to a rise in recycling demand. Given that it is produced in huge amounts at all stages of the food production and consumption chain, in both developed and developing nations, throughout the year, food waste is one of the key streams for waste avoidance/treatment globally. Food waste has grown to be an important issue in the world. In recent years, the world has faced a substantial threat brought on by the massive amount of waste that has been produced. Both rich and developing nations have suffered as a result. Food waste can account for up to one-third of all food produced for human consumption, with serious socioeconomic and environmental repercussions. The energy and time used to make the food are wasted when it is wasted/under-utilized. In addition to harming the environment, waste has enormous financial consequences. In a world where resources are becoming increasingly scarce, it is imperative to lessen the negative environmental, social, and economic effects connected with this type of waste.

An integrated system for waste management (ISWM) is defined by the Environmental Protection Agency (EPA) of the United States as a comprehensive waste reduction, collection, composting, recycling, and disposal system. To decrease, reuse, recycle, and manage the trash/waste, a strong integrated system for waste management framework is required to protect human welfare and the natural environment. In order to reduce trash/waste production and increase waste recycling, integrated waste surveillance systems evaluate/monitor effective waste gathering, transporting, and disposing together. Integrated waste management has emerged as a comprehensive tool to manage waste by combining and applying a number of acceptable methods, tools, and organizational plans in order to achieve specific objectives. The definition of integrated waste management has expanded beyond the classification of garbage to include a wide range of interconnected systems and objectives. The major goal is to determine a structure of orientation for developing and putting into use new waste management systems, as well as improving the ones that already exist. There is no perfect integrated waste management method that offers a solitary strategy suitable for managing all garbage/waste in a time-consuming manner.

Students, researchers, and scientists from all over the world will benefit from our insights into integrated waste management strategies.

– *Author*

INTRODUCTION

Villagers and cities are afflicted with a number of ailments/effects as a result of the inadequate hygienic conditions, despite the open landscape. The main causes of rural/urban pollution include improperly planned dwellings/houses and drainage systems, as well as the improper management of household, industrial, and human waste. The correct collection of waste, a suitable system for its transportation, and a safe, cost-effective disposal are fundamental requirements in rural areas if better sanitary conditions are to be ensured and the general health of those residing on farms is to be promoted. The effective waste management system will also aid in the efficient use of waste products for the production of electricity, raw materials for industries, and manure for agricultural areas and others. Since ancient times, disposing of waste in rural and urban areas has been a concern. In an effort to maintain and improve the environmental quality in light of industrial development and the constantly expanding populations of cities and villages, industrial pollution has been controlled and garbage/waste from both urban and rural areas has been disposed of. When it comes to rural management, proper trash/waste disposal is simply disregarded. To develop alternative methods and processes for the management and disposal of rural wastes, the issue/problem must be carefully considered given the availability of significant quantities of wastes from crops, animals, and human inhabitants.

Farm waste management and disposal cannot be done separately. Farm waste management can be characterised as a discipline concerned with the regulation of waste generation, storage, collection, transfer, and transport, processing, and disposal in a way that complies with best practices in public health, economics, engineering, conservation, aesthetics, and other environmental considerations. Farm waste management comprises all administrative, financial, legal, planning, and engineering tasks involved in the whole range of waste-related issues that residents of the community are forced to confront. Due to the number and variety of wastes, financing restrictions, the impact of technology, and a lack of social awareness within the rural community, agricultural waste management issues today are difficult. Resources that are not renewable have been significantly depleted as a result of food processing and packaging. In fact, daily needs for food processing, transportation, and preservation involve enormous amounts of water, air, power, and fuel consumption. The truth is that the environment is in a state of rapid decline, notwithstanding the agreements outlined in the Montreal and Kyoto treaties.

Chapter

1 INTRODUCTION TO INTEGRATED WASTE MANAGEMENT

Content

1.1 INTRODUCTION

One of the essential and fundamental needs for human survival is food. The greatest difficulty that humanity is currently facing is its waste. Approximately one-third of today's food production is wasted. Food waste is a financial and environmental issue/problem. On the one hand, there is massive food waste; on the other hand, one in nine of the world's population goes hungry every night, and about 21,000 people die every day as a result of malnutrition (Correia et al., 2014). Food waste (Figure 1.1) results in the loss of time, effort, and other resources used in the food producing process. According to reports, from 2005 to 2025, the amount of food wasted annually in Asian countries will increase from 278 to 416 million tonnes(Melikoglu, Lin, & Webb, 2013a). Approximately 28 percent of the agricultural area, or 1.4 billion hectares of productive land, is used each year to produce food that is wasted (Angus et al, 2009). The amount of waste produced is influenced by geography, culture, and eating habits. It is subjective to determine whether food is edible or not. Additionally, not all food or livestock products are edible.

Figure 1.1. Food waste.

(Source: https://upload.wikimedia.org/wikipedia/commons/thumb/3/31/GI_Market_food_waste.jpg/800px-GI_Market_food_waste.jpg?20130619040217, Creative Commons License).

Unavoidably, there will always be some inedible components like bones, fruit rind, eggshells, etc. A significant amount of food is also produced, processed, distributed, etc., along the food supply chain. Animals produce waste from food processing facilities and crops in the case of agriculture, in addition to the bulk of hazardous and toxic agricultural waste. Approximately 998 million tonnes of agricultural waste are thought to be produced annually. Farm organic wastes make up over 80% of all solid waste generation, and manure production accounts for 5.27 kg per day per 1,000 kg of weight (Obi, Ugwuishiwu, & Nwakaire, 2016). Reducing agricultural and food waste is seen as a way to lower production costs, enhance food safety and nutrition, and work towards the sustainable development goal. We must grasp the scope of the issue in order to avoid further confusion caused by the global level of food waste output. Food waste was not even acknowledged as a significant waste decades ago. Nevertheless, the attention has turned to food waste due to rising concerns about poverty and environmental issues. Wasted food raises environmental issues. Nearly 3.3 billion tonnes of CO_2 is emitted annually, according to FAO, which causes climate change and global warming (Swain et al, 2016). The decrease in food waste will also have a significant impact on other SDGs that are related to food, including SDGs 6 and 11 that deal with sustainable cities and communities, SDG 7 that deals with affordable and clean energy, SDG 9 that deals with infrastructure, industry, and innovation, SDG 13 that deals with climate change, SDG 14 that deals with marine resources, and SDG 15 that deals with terrestrial ecosystems, forests, land, and biodiversity.

Food waste results in the waste of resources used to produce food that is not eaten. In addition to being a waste of resources like energy, carbon, water, and nutrients, it has several negative outcomes. Food production necessitates irrigation water, which depletes groundwater aquifers and surface water bodies, having a negative effect on both the groundwater and surface water bodies. Food waste in junkyards and landfills can cause a number of ailments in the local community. The dumps also draw contagious disease carriers/vectors, that enter the food chain and increase the risk to human health. Another global impact of food waste is the loss of biodiversity as a result of land being converted from forests to farmland and the extensive ecological harm caused by food production. Food waste has prompted nations to unite and offer attractive international commitments, such as the "Paris Agreement," UNFCCC, which calls for coordinated action to reduce carbon emissions by limiting global warming to 1.5–2°C above pre-industrial levels by 2100, and the UN Sustainable Development Goal 12, "Responsible consumption and production," which includes an explicit goal of reducing food waste by 2030 (Rogelj et al., 2015). SDG 12 intends to eliminate food losses in production, postharvest, and the supply chain as well as to cut down on global per capita food waste.

1.2 AGRICULTURE WASTE

Agriculture waste, according to the United States Environmental Protection Agency, is a consequence of the cultivation, production, and harvest of crops and trees. Agriculture waste is defined by the Organization for Economic Cooperation and Development as waste generated from various agricultural processes. It includes farm waste, waste from slaughterhouses, and waste from runoff from fertilizers, pesticide residues, field silt, and salt.

Figure 1.2. Agricultural waste.

(Source: https://upload.wikimedia.org/wikipedia/commons/9/93/Farm_Waste_-_geograph.org.uk_-_1240814.jpg?20110226162833, Creative Commons License).

Agricultural waste, (Figure 1.2) broadly speaking, refers to the leftovers from agricultural products like fruits, vegetables, meat, poultry, dairy, and crops. The society can benefit from materials found in the agricultural waste. In contrast to the costs of collection, processing, and transportation, it is less important economically. The composition of agricultural waste varies depending on the agricultural activity. The waste can be a slurry, a liquid, or a solid. Untreated animal manure results in the production of GHGs, posing a major hazard to the environment. Additionally, the smell coming from the cages as a result of the animal waste's digestion process pollutes the air. Nearly 75–95 percent of the total volume of animal waste is made up of water, with the remainder being made up of organic and inorganic debris, as well as some kinds of microbes and parasites that can have a harmful impact on the environment and transfer disease to humans (Hamidian et al., 2021). Animal waste also includes waste from aquaculture. Most aquaculture waste is produced as a result of metabolism. Thirty percent of the material used as feed in aquaculture is wasted. A rise in temperature causes the feeding rate in aquaculture to increase, which in turn increases waste generation. Waste generation is impacted by water flow patterns as well.

1.3 FOOD PROCESSING WASTE

Food processing produces a tremendous quantity of waste for the food industry. The meat industry generates a large amount of solid and liquid waste, including skin, feces, blood, urine, etc., as well as foul odors. The smoke that is produced as a result of several processes, such as the evaporative/steam process that generates a lot of smoke, also contributes to air pollution. Almost no solid waste is produced by the milk processing industry. However, this industry produces a lot of wastewater. Sludge created during wastewater treatment is the only solid waste that is generated by this sector. The cleaning and sanitation processes generate a lot of wastewater. Cleaning of tanks, storage trucks, and in-line pipe washing and sanitizing all result in the production of wastewater. The formation of curd particles, whey, and buttermilk arises from the enormous amount of wastewater that is also produced when making curd, cheese, and butter.

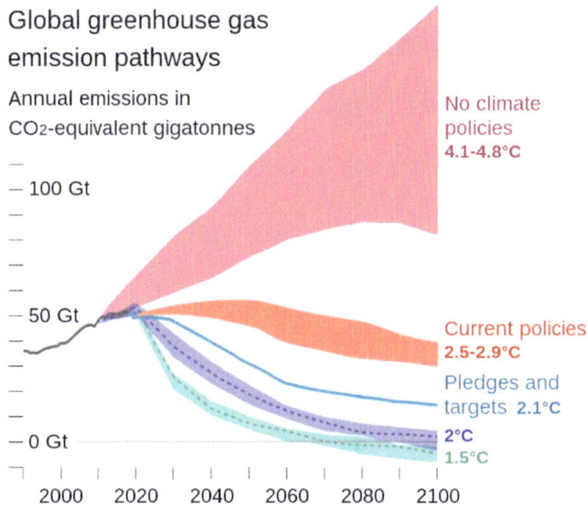

Figure 1.3. Greenhouse Gas (GHG).

(Source: https://upload.wikimedia.org/wikipedia/commons/thumb/6/6e/Greenhouse_gas_emission_scenarios_01.svg/705px-Greenhouse_gas_emission_scenarios_01.svg.png?20220808215131, Creative Commons License).

Milk production has increased by about 2.8 percent annually as a result of industrialization, and the processing of dairy products is thought to be a significant contributor to industrial effluent (Zhang et al., 2017). In addition to this, the dairy industry adds to air pollution by releasing GHGs (Figure 1.3) like CO_2, CO, and SO_2.

Due to the usage of a vast variety of fruits and vegetables, the wide range of procedures, and the diversity of the product, the waste produced by the fruit and vegetable processing sector varies greatly. This industry produces a significant amount of waste in the form of fruit peel and skin. Since fruit skins are typically removed before consumption, there is a greater amount of waste produced. Between 25 and 30 percent of the vegetables and fruit are inedible. Pods, peel, and seeds leftovers are produced in significant quantities during the preparation of fruits and vegetables. The fruit and vegetable industries produce a lot of waste, including pomace from the grape and apple industries, peel from the banana and citrus industries, seed from the mango and guava industries, and shells from the pea sector. The main components of the processing waste are fiber and soluble carbohydrates. Environmental issues arise from the landfilling of the waste. Fruit washing and equipment cleaning and sanitization produce the liquid waste from this industry, or wastewater. Liquid waste disposal introduces extra nutrients to the aquatic environment. Crustaceans, amphibians, and fish perish as a result of the unbalanced nutritional balance, overgrowth of algae, and decrease in oxygen levels. Field Waste comprises waste left behind from harvesting in fields or orchards. It is made up of things like leaves, twigs, stalks, and pods. This leftover material is either burned or plowed back into the ground. The act of igniting agricultural residue to remove it from the field for the planting of the following crop, has a variety of negative effects. Burning causes the emission of gases into the atmosphere such as CH_4, CO, volatile organic compounds, and polycyclic carcinogenic aromatic hydrocarbons, which create a heavy, thick layer of smog that is harmful to human health and adds to air pollution (Dutta et al., 2017). In addition, the heat reduces soil fertility by nitrogen depletion and causes the loss of beneficial bacteria.

1.4 PROCESSING WASTE

Process waste includes elements discarded after the crop is converted into a useful resource. It includes things like bagasse, molasses, seeds, husks, and roots. This waste can be used to improve soil fertility and as animal feed. The development of weeds and insects is also aided by the growth of the crop.

The producers encourage the use of a variety of insecticides and pesticides to address this issue. After using pesticides, they are dumped directly into the fields or ponds, posing substantial environmental issues like contaminating farmland. The presence of pesticide residue in crops also contributes to the problems with food safety and food poisoning. Some farmers employ the needless use of cheap inorganic fertilizers to increase the output of their produce. When fertilizer is used excessively, part of it is absorbed and kept in the soil, and some of it is washed into lakes and rivers by irrigation or runoff. By evaporating, the sprayed pesticides and insecticides also enter the environment, contributing to air pollution. The requirement for food has significantly expanded along with population growth, urbanization, and economic growth, which has boosted food production and consequently increased waste output. Urbanization is a byproduct of industrialization, which is currently underway in emerging nations like India. The population shift to cities has increased the amount of solid waste produced per person. This suggests that a significant portion of the resources utilized in food production are wasted. Food is wasted in large quantities throughout the supply chain. There are 2.01 billion tonnes of municipal solid refuse produced each year, and roughly 33% of it is not handled in a way that is safe for the environment (Roy et al., 2022). East Asia and the Pacific generate around 23% of the world's waste, while the Middle East and North Africa produce 6% (Sadiq et al, 2023). The fastest-growing regions include South Asia, the Middle East, North Africa, and Sub-Saharan Africa, where waste creation is expected to quadruple and triple, respectively, by 2050 (Ahut Daggash & Dowell, 2021).

The need for agrarian food is increasing due to the world's rapidly growing population, which has increased from 2.50 billion in the 1950s to 6.90 billion in 2010 and may reach 9.15 billion in the 2050s (Ghose, 2014). About one-third of the wastes from the agri-food industry is converted to post-utilization wastes, which include cooked meals and leftovers on consumer plates, during the pre-utilization process. Agri-food wastes are estimated to amount to 1.30 billion tonnes annually, according to the Swedish Institute for Food and Biotechnology (Kosseva, 2009). Additionally, the FAO of the United Nations estimated that in 2009, around 32% of all the food and consumables transported around the world were lost or wasted (Kummu et al., 2012). It is estimated that 870 million people worldwide suffer from acute undernourishment or starvation. This equates to nearly one-third of the 1.30 billion tonnes of food produced each year for human consumption. Byproducts and waste are produced in greater amounts as a result of agricultural output and agro-industrial processing techniques. More than half of all fresh fruits contain leftovers like cuttings, peels, shells, stems, and seeds that may be more nutritious or useful than the final product. Food waste is also a result of the damage that is done during transportation, storage, and handling.

According to reports, the creation of byproducts and wastes has increased recently due to the growing popularity of fruit juices, nectars, solidified foods, and minimally prepared foodstuffs.

The horticulture and agro-industrial divisions deliver enormous amounts of residues or derivatives that are made up of various natural mixtures, including animal waste, food processing waste, harvest waste, and dangerous and lethal agricultural waste. There are wastes from food hubs, eateries, and bars, as well as wastes from food generating services, which primarily consist of used cooking oil, leftovers, and perished or subpar food. Around the world, more than 1 billion tonnes of food waste are created each year, the bulk of which might potentially be recycled (C. Liu et al., 2016).

Large-scale deterioration of agricultural, agro-industrial, and food wastes is the cause of pollution of the land, water, and air. Diverse greenhouse gases released from solid waste disposal sites also contribute to global warming. These reasons for severe pollution issues are focusing attention on tightening environmental regulations in almost every nation. Since food processing methods involve many unit operations, such as washing, evaporation, removal, and filtering, the wastewater is typically the main cause for concern. The process of wastewater operations increasingly frequently includes high quantities of soluble organics such as proteins, lipids, and carbohydrates as well as suspended particulates, which are referred to as challenging disposal problems. The standard methods of composting, discarding, burning, and anaerobic absorption are used to dispose of food waste (FW). Due to the high moisture content of FW methods, for example, combustion produces harmful fumes, and composting results in the release of leachates, making the process dangerous to the environment. Conventional techniques of garbage disposal are not suitable for food waste since they need a lot of energy, and produce dangerous gases, leachates, and odors, which pose a threat to human survival.

After being properly utilized, waste from foods and their derivatives can be used as raw materials or added to foods as additives, resulting in the financial expansion of businesses, a reduction in the food crisis, the emergence of positive well-being as a result, and a reduction in the ecological effects that result in the disorganization of waste. Industries are currently using new technology to attain zero waste, allowing the waste produced to be used as raw material for the creation of novel goods. Scientists are continually looking for alternatives to handle solid waste to protect the environment. Reduced food and agricultural wastes and their usage in valued products have the potential to strengthen the food supply chains (FSC) (Garcia-Herrero et al., 2018).

Reducing resource use and recycling are higher on the waste management food chain since they are more environmentally friendly. Basic reduction starts with reducing the amount of waste created and reusing things to prevent them from entering the waste stream. The concept of reducing waste decreases the severity and negative effects of waste production by reducing wastes, recycling waste with simple refinements, and rescuing waste by treating it as a resource to create the same or tailored products. This is commonly referred to as "3R." By employing the same resource to produce different or the same object, a small amount of waste might be consumed.

1.5 CROP RESIDUES

Crop remnants have a high fiber content but a low protein, carbohydrate, and fat content. As a result, the traditional method of increasing livestock production by supplementing the feed and fields with grains and protein concentrates may not be able to satisfy the projected needs for meat protein. Utilizing protein and grains for human consumption will compete with this usage of fodder.

These scenarios might be avoided by feeding domesticated animals leftovers. Anaerobic digestion and landfilling are not significantly more advantageous than feeding animals under well-managed settings.

Benefits mostly result from the partial imitation of regular feed, whose production has a significant negative influence on the environment and human health. Local legislations in nations like Japan and South Korea encourage the use of AFW to feed animals, inflating the portions of food waste recycled as forage to 35.90% and 42.50%, respectively (Coudard, Corbin, de Koning, Tukker, & Mogollón, 2021). Although it is now illegal to use food waste as fodder in the EU, these applications are clearly lawful there.

Figure 1.4. Landfill and agricultural waste.

(Source: https://upload.wikimedia.org/wikipedia/commons/thumb/1/13/Tagebau_J%C3%A4nschwalde_Entw%C3%A4sserungsbrunnen.JPG/800px-Tagebau_J%C3%A4nschwalde_Entw%C3%A4sserungsbrunnen.JPG?20151031204704, Creative Commons License).

When carbon and nutrients enter landfills (Figure 1.4) as agricultural waste, they can build up inside the dump or are released as vapors and fluids. The majority of the energy is used for the collecting of waste, its transportation to landfills, and the transportation of leachate to remote locations for treatment in accordance with regulatory dispensing requirements. After discharge, contaminants become concentrated and may exceed the ambient water quality limits for surface or groundwater. As a result, it should be diluted with a lot of water to bring it back within the acceptable ecological standards. The biggest human source of atmospheric CH_4 emissions, landfill methane accounts for around 800 million tonnes of CO_2-equivalent and is created by the anaerobic biological decomposition of organic waste(Pour, Webley, & Cook, 2017). In wealthy nations, the collecting rate of landfill gas is often less than 60%. The US EPA estimated that 282.6 million tonnes of CH_4 were produced by human activity in 2000, of which 13% came from landfill emissions (X. Liu et al., 2012). AD or anaerobic digestion has been effectively used and recognized as a method for disposing of food and agricultural waste in the treatment of industrial and agricultural waste. Since

2006, this method has been widely used for waste treatment in the European Union and wealthy Asian nations. This has become a well-known technology used to handle manure created in wastewater treatment plants. AD is used to minimize food and agricultural waste, which entails degrading waste in the absence of oxygen. Methane, ammonia, and carbon dioxide biogas, along with an aqueous product like fertilizers, are the products acquired during AD.

Hydrolysis is the process by which carbs, proteins, amino acids, long-chain fatty acids, and fats are converted into sugar, respectively, and polymer organic material in solids is initially broken down into soluble smaller units. Because volatile fatty acids or hazardous byproducts are produced when the raw substance is degraded, the majority of researchers have suggested that the hydrolysis level is the rate-limiting stage. The subsequent process, known as acidogenesis, involves monomers degrading into short-chain fatty acids together with volatile fatty acids (VFAs), including formic acid, pyruvic acid, and acetic acid. The digestion of acids like lactic acid and pyruvic acid into acetic acid and hydrogen then takes place during the process of acetogenesis. Methanogenesis is the process by which methanogenic archaea transform hydrogen and acetic acids into CH_4 in the final stage. Methane makes up 50 to 70 percent of the average gas produced, along with CO_2, 25 to 45 percent, N_2, 0.5 to 3 percent, and H_2 with traces of H_2S (Bhawangirkar & Sangwai, 2019). The amount of energy required to heat the gas is in the range of 18 to 25 MJ/m^3. AD enables the processing and disposal of massive poultry, swine, and dairy waste, hence reducing the odor issue. The waste is neutralized, the digesting sludge is comparatively odorless, and the original waste's fertilizer value is still preserved.

Because the process can transform this unique biomass resource into non-condensable gases, and pyrolysis oils, which could produce useful energy and chemical goods due to their high calorific value, pyrolysis has been seen as an attractive alternative to dumping agrarian waste. A portion of agricultural waste is heated in pyrolysis systems to 400–600 °C in the absence of oxygen, vaporizing the material and leaving char behind. This is thought to be a cutting-edge technological method for using agricultural wastes. They are employed in energy recovery as well as the production of compounds from agricultural waste. Oil-derived biomass may be utilized as a feedstock for bio-refineries or current petroleum refineries to make hydrocarbons that could be mixed easily. The gas that is produced can be utilized to make liquid fuels or as a fuel source for the creation of synthetic natural gas. The charcoal can be used as a soil fertilizer, as a reducing agent in the metallurgical industry, as a fuel in homes, as an adsorbent or as a raw material for the synthesis of chemical compounds like calcium carbide, silicon carbide, sodium cyanide, carbon black, etc. Even though several projects have been approved, pyrolysis of biomass is still in the process of being developed for marketable usage and must overcome numerous financial and technological obstacles to compete with traditional fossil fuel-based processes.

Because of the activities of microbes like bacteria, fungi, and protozoa, organic materials undergo aerobic digestion into rather predictable products. The organic matter is converted into CO_2, NH_3-N, or complex insoluble compounds, which are usually referred to as humic under aerobic conditions. Compost can be used as a soil repair, which eliminates the production and use of chemical fertilizers, captures C, promotes soil moisture retention, and reduces the need for irrigation.

Figure 1.5. Vermicomposting.

(Source: https://upload.wikimedia.org/wikipedia/commons/thumb/c/c8/Vermicomposting_of_toilet_blackwater_1.jpg/528px-Vermicomposting_of_toilet_blackwater_1.jpg?20161222213317, Creative Commons License).

More recently, vermicomposting (Figure 1.5) of food waste has been developed for the production of bioinoculants, which embodies the stability of organic waste due to the synergism (symbiosis) between earthworms and bacteria. The finished product has excellent soil nutrient availability and is regarded to have superior qualities to its conventional equivalent.

Vermicomposting is a bio-process that uses unprocessed waste as a substrate to mineralize it and turn it into nutrient-rich unprocessed dung. Vermicompost can be used to improve soil, and earthworms are used in various technologies including vermin-filtration and vermi-remediation as well as in medicine and as food (Emamjomeh et al, 2018). Vermicompost is frequently used in chemical-free agriculture as raw fertilizer. Therefore, vermicomposting is preferable to artificial fertilizers and is good for crops. Its technique allows for the supervision of unprocessed waste, making it suitable for use with food and agricultural waste.

As many techniques have been utilized to stabilize waste produced as a result of various agronomic activities as well as from other businesses, vermicomposting has been adopted as one of the non-polluting technologies for waste organization. Vermicomposting produces products including vermicompost, and vermiwash. Vermicomposting has also been employed as a practical, affordable method of handling waste water. The administration of sewage water treatments is aided by vermi filtration. Vermicompost is a suitable organic adsorbent that is inexpensive because of its large surface area and increased cation action. Therefore, vermicompost is also used to remove colors and heavy metals from wastewater. According to research, vermicomposting treatment also yields a liquid byproduct that is environmentally friendly in addition to vermicompost. Support is given to the chemical-free process of preparing these liquids as a means of creating environmentally acceptable biofertilizer. Because of the way vermicompost is processed, one advantage of the liquids produced from it is that all of the nutrients are completely soluble in water. It could therefore be applied as a foliar fertilizer. Vermiwash, vermicomposting leachate, and vermicompost aqueous extract were the subjects of experiments, and the results showed that these biofertilizers had a guaranteed impact at different dilutions. Vermicomposting made from liquids hence has the potential to be applied as foliar fertilizer. Food waste has a high moisture content, which increases the amount of energy needed for combustion and ignition. As a result, agrifood waste or AFW is typically burned where other inflammable waste provides energy. The heat created during incineration is used to produce electricity. The rapid chemical reaction of biomass and oxygen, the release of energy, and the quick development of the final oxidation yield of organic matter- CO_2 and water- are all components of absolute combustion of agro waste. If the oxidation proceeds at an acceptable rate, the energy normally dissipates in the form of radiant and thermal energy; the amount of which is enthalpy of burning the biomass.

Worldwide, more than 1 billion tonnes of food waste are produced each year, the bulk of which may be recyclable. It is interesting to note that wastes processed from the agri-food industry include a loss of valuable biomass and other supplements because these wastes have the potential to be enhanced into beneficial goods or even used as raw materials in a variety of businesses. The company might potentially grow economically if agri-food waste and byproducts are used properly as raw materials or ingredients for food. This will also reduce health issues, improve positive health effects, and lessen the environmental load that results from improper waste management. Today's companies are concentrating on formulas to produce zero waste, where the garbage created by one might be used as the raw material by the other to create new products. As an environmentally friendly fuel alternative to crude oil, biodiesel, also known as a monoalkyl ester of fatty acids or fatty acid methyl ester, is a growing industry. It is now one of the best gasoline alternatives available. In the EU and the UK, fats are used to create biofuel.

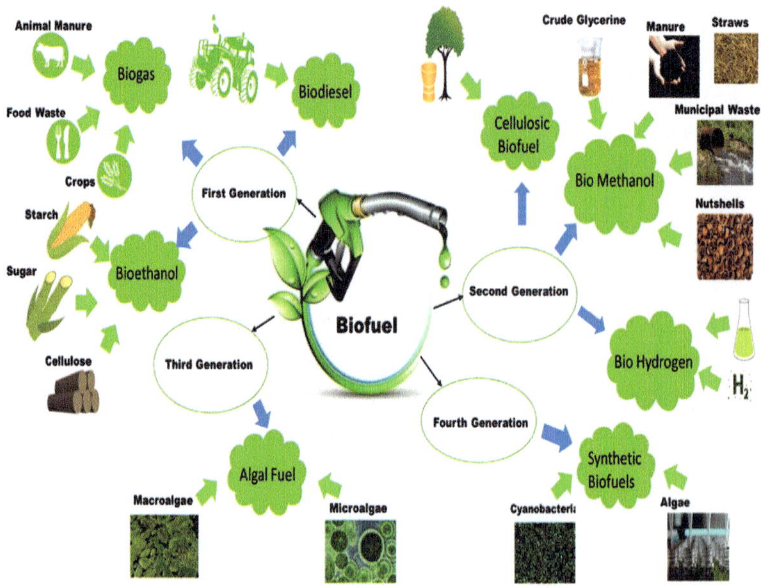

Figure 1.6. Biofuel.

(Source: https://upload.wikimedia.org/wikipedia/commons/thumb/d/d9/Types_and_generation_of_biofuels.png/800px-Types_and_generation_of_biofuels.png?20201010040502, Creative Commons License).

Biofuel (Figure 1.6) can be made from food waste by combining the fermentation of microalgae and the transesterification of enzymes (Lin et al., 2014).

1.6 WASTE OIL

The recent development of revolutionary technologies has made it easier to produce biofuel from used frying oils, leading to the development of a product whose quality is comparable to that of vegetable oil biodiesel. Pre-processing is necessary for used cooking oil in order to remove solid impurities and reduce the level of free fatty acids or FFAs present. Centrifugation, washing, acid esterification, and flash evaporation are pre-treatment procedures.

Waste lipids can be combined and diluted with petro-diesel for direct use; nevertheless, the process of trans-esterifying the lipids into biodiesel is a crucial one where a lot of activities take place. Alkaline, acid, or enzymes can all catalyze the transesterification reaction. Food wastes, including bone, ash, and shell, can be processed and used as a catalyst to aid in the release of calcium oxide. Pretreatment of oil-linked acid esterification is necessary to reduce the level of working reactants required for saponification and to reduce catalyst inhibition, but the pretreatment phase must be tuned due to the varying composition of various feedstocks. Utilizing innovative technologies like ultrasound and microwave can assist the transesterification process and increase product yield. The structure needs to be dry during the transesterification process to prevent saponification.

Figure 1.7. Cooked oil wastes.

(Source: https://upload.wikimedia.org/wikipedia/commons/thumb/3/39/Waste_Restaurant_Cooking_Oil_Truck_in_Hempsted_New_York.

jpg/800px-Waste_Restaurant_Cooking_Oil_Truck_in_Hempsted_New_York.jpg?20220326174420, Creative Commons License).

Cooked oil wastes, (Figure 1.7) in addition to canola and soybean oil, are the raw materials used to make biodiesel. A byproduct of the palm oil industry can be used as a feedstock to manufacture biofuel with little effort. In addition to lipases, microbial enzymes are used in the production of biofuel. For the production of biofuel from soybean oil, mixed lipases from *Rhizopus oryzae* (Figure 1.8) and *Candida rugosa* are used to damage the silica-gel matrix (Abdulmalek, Li, Wang, Ghide, & Yan, 2021).

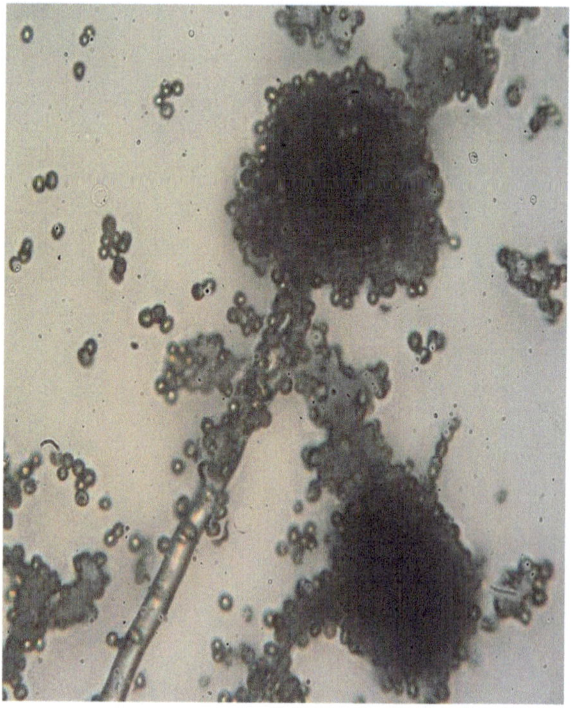

Figure 1.8. *Rhizopus oryzae*.

(Source: https://upload.wikimedia.org/wikipedia/commons/thumb/c/c5/Rhizopus_fungus.jpg/625px-Rhizopus_fungus.jpg?20100729140737, Creative Commons License).

Using this process, elevated biodiesel levels were achieved, and the surrounding material could be reused up to 30 times. As a result of the transesterification of microbial oils produced by various oil-producing bacteria, food waste can be converted into fatty acids and biodiesel. In this instance, the food waste serves as a substrate for the culture of oil-producing microorganisms to accumulate oil with or as a result of transesterification. Due to their similar fatty acid composition, oils produced by microorganisms can be used as a substitute for oils derived from plants. These oils can be produced by various strains of yeast or microalgae. Phosphorus and glucose were recovered from food wastes by hydrolyzing the fungi *Aspergillus oryzae* and *Aspergillus awamori*, and these nutrients can be used as culture medium for the two heterotrophic microalgae *Chlorella pyrenoidosa* and *Schizochytrium mangrovei*. Using FWH (food waste hydrolysate) to produce 10–20 gm of biomass allowed for the fine development of *C. pyrenoidosa* and *S. mangrovei*. The two biomass strains were said to be suitable for producing biofuel since they were rich in SFA (saturated fatty acids).

Due to the residual oil present in the agro-industrial wastes and agriculturally processed wastes including used vegetable oil, rice bran, and deoiled cakes of both edible and non-edible oil seeds are potential feedstocks for the production of biofuels. About 482 million metric tonnes of rice will yield around 38.50 MMT of rice bran (Giwa, Alabi, Yusuf, & Olukan, 2017). The approximate oil content of rice bran is 17.50%, and the crude rice bran oil contains 8% FFA, which has tremendous potential for the generation of biodiesel. Due to the presence of a high free fatty acid content, two-step pretreatment techniques are used to make rice bran oil.

Acid-catalyzed esterification first occurs in the presence of sulfuric acid and methanol, resulting in the production of low-quality rice bran oil and transesterification. The amount of biofuel generated by this process, which was observed, varied from 60 to 85 percent depending on the methanol percentage, catalyst percentage, and reaction time. Solid acid catalysts, which offer advantages over mineral acids or base catalysts due to their variety of characteristics, can also be used to produce biofuel. Non-edible oil seeds including karanja, jatropha, simarouba, cottonseed, neem, etc. have mostly gained recognition as practical sources of materials for biofuel production. However, the method used to extract oil from the aforementioned seeds has a significant impact on the production of biofuel. Despite the fact that numerous researchers have developed a variety of effective oil removal techniques from inedible seeds, not a single oil extraction method has been produced. Less oil in the cake makes it suited for the production of biological CNG with added value like amino acids and proteins. The economy of biofuels is positively impacted by the wise usage of deoiled cake.

1.7 PHENOLS

Phenolic complexes are often found in the kingdom Plantae and are made up of a broad range of secondary metabolites derived from phenylalanine and tyrosine. Phenolic complexes can be categorized into many categories, such as C_6H_6O, $C_9H_8O_3$, $C_7H_6O_3$, $C_8H_8O_2$, and more phenylpropanoids, as well as HT and CT, depending on their basic carbon structures. These substances can be arranged in a variety of ways, from simple, large polymer masses to complicated, elementary phenolic molecules, all of which have aromatic loops holding one or more hydroxyl clusters. The arrangement of phenolic complexes, particularly the type of substitutions on aromatic rings and the number and locations of the hydroxyl groups, affects their antioxidant activity. Fruits, vegetables, and beverages make up the bulk of a man's diet and are sources of phenolic chemicals. Phenolic compounds have a wide range of physical properties, including effects that are anti-thrombotic, anti-microbial, anti-allergenic, anti-inflammatory, anti-atherogenic, antioxidant, and vasodilatory. Therefore, eating plant-based foods has a positive impact on health and a reduction in disease. The demand for natural products, which are said to enhance health while also being safe, is also growing as customers increasingly become aware of health-related issues. The production, arrangement, and use of food generates substantial amounts of solid and liquid waste in the agriculture-food industry. The food industry uses a lot of water, and some of the water may end up in the final products. The byproducts from the agrarian-food industry are frequently an inexpensive source for acquiring bioactive chemicals as well as high phenolic goods. Due to the recent expansion of the agricultural and food industries, numerous studies have been conducted to identify the presence of phenols and determine their antioxidant capacity in various byproducts/leftovers.

Because of the rise in demand and consumption of natural byproducts in recent years, the processing of fruits and vegetables has significantly expanded. There is a significant amount of fruit and vegetable waste produced by different food sectors and other wholesale markets. A significant portion of agro wastes consists of unprocessed fruits and vegetables like tomato, mango, pineapple, jackfruit, orange, and banana. Fruit and vegetable wastes present a challenge for landfill disposal due to their perishability and environmental risks. The yearly production of fruit and vegetable waste in India is 5.60 MT, and at the moment, the waste is dumped outside of cities (Kumar & Sharma, 2014). Concentrated wastewater is formed during food preparation methods such as bleaching vegetables. Water from canneries is comparable to residential kitchen waste. The trimming, cutting, bleaching, and juice processing of fruits and vegetables results in the creation of waste. Large levels of dissolved and suspended particles are present in the waste water, with fruit sugars and starches making up the majority of the combination. For example, the waste produced by pineapple canneries includes between 85 and 90 percent sugar (Domínguez-Bocanegra, Torres-Muñoz, & López, 2015). Huge amounts of fermentable sugar are present in the fruit and vegetable waste, and peas, apples, oranges, and carrots have all been successfully employed as a substrate for fermentation. Vinegar, acetic acid, and citric acid have all been made from the waste products of the processing of fruits and vegetables.

95 percent of the sugar beet waste is made up of sugar, with the remaining 5 percent being raffinose (Cieciura-Włoch, Borowski, & Otlewska, 2020). Nitrogen and phosphorus levels in this refuse are minimal. These sugars are converted into short-chain aliphatic carboxylic acid, leaving just acids in the wastewater that need to be treated. However under acidic conditions, sulfides and VFAs may cause offensive scents. An appropriate amount of lime should be added in order to keep the pH of the circulating water neutral. Since strawberries are a seasonal crop, their extreme perishability endangers their quality, causes significant loss, and has a negative impact on the economy and ecology. Due to the chemical composition of strawberries, they can be used to create new goods. For instance, strawberry puree can undergo biotransformation, which results in products with higher quality and health advantages. *Gluconobacter* sp. can use the acid for other biotransformations, but the strawberry puree's glucose can still change into gluconic acid. Bioconversion can be accomplished without using extra fructose to make the final product sweet. The finished product can be used to make a variety of drinks and condiments by combining it with vinegar in different amounts.

One of the foods that people consume frequently is apples. These are a top source of phenolic acids including quercetin, epicatechin, catechin, and naringenin-like compounds as well as $C_{21}H_{24}O_{10}$ and antioxidants. Industrial processing creates tonnes of byproducts each year, the majority of which are juices and frozen, and dried vegetables. Apple pomace, which represents up to 30% of native fruit, is the refuse/waste created by the subsequent manufacture of apple juice. It mostly consists of the peels/flesh, stems, and seeds. Apple leftovers that have just been crushed should be dried out quickly to prevent microbial deterioration caused by the high moisture level. Apple pomace contains 35–60% nutritional fiber along with other substances like tannins, resins, reducing sugars, and pigments that are referred to as extractives because they dissolve in water or organic solvents. Successful efforts are being made to use apple residue as a source of dietary fiber in yogurt, meat products, and bakery goods. Researchers found that apple pomace and oat bran can help stabilize oil-water phases. A notable and characteristic apple polyphenol known as phloridzin, a chalcone derivative, guards the apple plant against various bacteria. Apple pomaces also have an anti-diabetic effect because they inhibit the Na-linked $C_6H_{12}O_6$ carriers and interfere with the absorption of glucose. Due to its antioxidant properties, it can stop lipid peroxidation, stop bone loss, improve memory, and even stop the growth of malignant cells.

Ascorbic acid, lupeol, and beta-carotene, three polyphenols, are highly concentrated in mango. Since mangoes are a seasonal commodity, around 20% of the fruit is converted into products including puree, canned slices, nectar, pickles, leather, and chutney (Evans, Ballen, & Siddiq, 2017). The byproducts and seeds that are created when handling mangoes make up about 13–16 percent. Mango peels make up roughly 15 to 20% of their overall weight. Mango peels (Figure 1.9) have a significant amount of soluble and insoluble dietary fibers (225-725 mg/g), as well as roughly 100 mg/g of phenolic compounds, which include hydrolyzable tannins, flavonoids, alkylresorcinols, and proanthocyanidins. The type, stage of maturity/ripeness, and processing methods all have an impact on the qualitative and quantitative composition of phenols.

Figure 1.9. Mango peels.

(Source: https://upload.wikimedia.org/wikipedia/commons/thumb/1/1c/Mango_peel_2.jpg/800px-Mango_peel_2.jpg?20170628172515, Creative Commons License).

Figure 1.10. Mango seeds.

(Source: https://upload.wikimedia.org/wikipedia/commons/thumb/2/2d/Seeds_of_Mango_tree.jpg/800px-Seeds_of_Mango_tree.jpg?20160513172246?20220326174420, Creative Commons License).

Mango seeds (Figure 1.10) have a high level of antioxidant properties and are rich in phenols. It is interesting to note that the characteristics of phenols in mango peels differ from those discovered in seeds from the same species of mango, with gallates and GT (gallo-tannins) being the dominant sets in peels and $C_{14}H_6O_8$ and its precursors in seeds. With these intriguing characteristics, studies have shown that powders made from mango peel and seeds are a promising source of functional food ingredients that may be added to foods like macaroni and biscuits to increase their potential for bioceutical benefits. Mango byproducts can also be used in the production of useful packaging materials. For instance, adding mango peel extracts to gelatin-based coatings for active food packaging improves coating strength and has good antioxidant activity. Mango extracts and constituent parts have been found to have qualities that protect both humans and animals from inflammation, oxidation, obesity, heart disease, diabetes, and skin cancer while also protecting the neurological system and gastrointestinal health. Mango peel pectin has the potential to be used in the food and pharmaceutical industries. Pectin has attracted the interest of the scientific community because it plays a significant role in preventing and reducing the likelihood of malignancies.

1.8 FOOD PRODUCTS AND USES

The root vegetable and a member of the Apiaceae family, the carrot can be consumed raw, cooked, or juiced. In addition to having a high anthocyanin, carotenoid, and fiber content, carrots also have a high antioxidant profile. Furthermore, carrots have anti-inflammatory, anti-cancer, and antioxidant chemicals that promote cardiovascular health and slow the onset of degenerative diseases. In addition, grated carrots can be added to cakes and puddings. Mini or baby carrots are sold in various stores and are eaten as a prepared snack. Carrot juice is the most popular health beverage due to its outstanding health benefits and diverse nutritional qualities. Peels and pomace, which are produced in significant numbers during all of these processes and are an excellent source of several useful elements, particularly bioactive compounds with antioxidant capabilities, are produced. It has been noted that carrot peels are a good source of different phenolic compounds. $C_{16}H_{18}O_9$ is the most prevalent phenolic acid found in carrots, followed by hydroxycinnamic acids and its prototypes such as 4-p-coumaroylquinic acid, caffeic acid, 3-caffeoylquinic acid, 3,5-dicaffeoylquinic acid, and 3,4-dicaffeoylquinic acid(Sun et al., 2017).

Figure 1.11. Carrots.

(Source: https://upload.wikimedia.org/wikipedia/commons/thumb/c/cf/Carrots_-_Daucus_carota_subsp._sativus.jpg/800px-Carrots_-_Daucus_carota_subsp._sativus.jpg?20190909035700, Creative Commons License).

Carrots (Figure 1.11) contain isocoumarins 6-hydroxymellein and 6-methoxymellein. The amount of phenolic compounds was examined in carrot peels under different working conditions, and the total phenolic levels discovered were around 1,380 milligrams/100-gram dry wt. Carrot skin contains pectin, a soluble dietary fiber made up of D-galacturonic acid, L-arabinose, L-rhamnose, and D-galactose units as well as other monosaccharides, and it is a vital part of the vegetable. Jellies, jams, frozen foods, and low-calorie foods are examples of functional foods that can use pectin as an additive or a complement. Pectin, an acidic structural heteropolysaccharide with thickening and stabilizing capabilities, has a lot of potential to replace hydrocolloids.

Fiber-rich carrot-enriched cookies were discovered to have improved color characteristics, better resistance to water absorption, and reduced softening. The dough and its growth also demonstrated increased stability. Results also indicated that fibers have high levels of antioxidant and prebiotic properties. The first vegetable deemed safe for use in the biopharmaceutical industry was a carrot. Consuming carrots, carrot byproducts, or goods enhanced with carrots has positive effects on a number of conditions, including hypertension, constipation, cardiac issues, cancer, infections, atherosclerosis, bronchial asthma, blood sugar regulation, and preventing muscular degeneration. Carrot peels are a great source of polyphenols, which have anticancer and anti-mutagenic properties and shield cells from harmful free radicals.

The most important vegetable crop, the onion is used in a wide variety of dishes and is now available in fresh, chopped, frozen, powdered, pickled, canned, and dried forms. The majority of the onion byproducts are the small, malformed, or ugly onion bulbs as well as the outer papery brown, red, or yellow skins produced during mechanical peeling of the roots and bulb. The thick, white onion scales are a great source of phenolics and organosulfur compounds. Alkenyl cysteine sulfoxides, fructans, flavonoids, and dietary fiber are all abundant in onion waste and are helpful for human rehabilitation. There is a total of 16 distinct flavonols found in onions, including aglycones and glycosylated quercetin prototypes (Wang, Li, & Bi, 2018). These flavonol chemicals make up roughly 90% of the onion bulb's internal and external scales. They are quercetin 3-glucoside, quercetin 3,4-diglucoside, isorhamnetin 4-glucoside, and quercetin 4-glucoside. The main consequence of onions, which are high in quercetin glycoside, is dry peel, which is also an oxidative byproduct. These substances are strong antioxidants that offer protection against cellular damage and are highly reactive against damage brought on by oxidative stress. Dried onion peels are a good source of ferulic, kaempferol, protocatechuic acids, and gallic in addition to quercetin compounds. Red onions had a total phenolic content of 105.50 gm per kg dry weight (Olsson, Gustavsson, & Vågen, 2010). The onion leaf extract was found to be a reliable natural colorant and presents a possible alternative to artificial colorants.

Quercetin has been demonstrated to have potential anti-allergic, anti-cancer, anti-asthma, anti-coronary heart disease, anti-arthritis, anti-osteoporosis, anti-neurodegenerative, and anti-gout properties. The cosmetic and pharmaceutical sectors have the ability to make use of the enormously helpful bioactive phytochemicals found in onion waste and byproducts. The anti-diabetic, anti-cholesterol, anti-carcinogenic, anti-inflammatory, antifungal, anti-microbial, antioxidant, anti-osteoporosis, and hypotensive effects of onions are related to each other. The majority of waste in the meat industry is produced during the slaughtering process. Only one-third of an animal after slaughter and processing is meat; the remainder consists of byproducts and waste. As a result, the waste produced must be treated correctly before being used effectively. Apart from the dressed carcass, every component of a slaughtered animal is a byproduct. Skin, tendons, bones, pieces of the GIT, blood, and internal organs are among the waste products in the meat business. Different types of animals produce different wastes. The optimal use of meat byproducts by the meat industries is essential for profitability. Animal waste is divided into edible and inedible byproducts based on whether it may be used as food or not. Humans can consume the brain, liver, spleen, kidney, heart, tongue, intestine, and other organs that are edible byproducts. On the other hand, byproducts that humans cannot eat as food are known as inedible byproducts. Examples include skins, hair, ears, gallbladders, fetuses, horns, bristles, etc. The term condemned meat also includes the organs of dead animals. Edible meat byproducts like liver and kidney have a higher carbohydrate content than other meat components and contain a variety of essential nutrients.

Byproducts have a different amino acid composition than lean tissues because they contain a lot of connective tissue. Therefore, byproducts like feet, lungs, ears, and stomach contain a significant amount of glycine, $C_5H_9NO_3$, and proline, while tryptophan and tyrosine are present in much less amounts. Compared to lean meat tissue, organ meat typically contains more vitamins. Animal blood, which has a high concentration of protein and heme iron, is a significant food byproduct. In Europe, several foods prepared from animal blood include blood pudding, blood sausages, biscuits, and bread. It is used in blood curd, blood pudding, and blood cake in Asia. Additionally, it is used to make non-food products including fertilizer, binders, and feeds.

Due to its useful qualities and lack of color, plasma, a component of blood, is extremely useful. In addition, it has excellent foam-forming abilities and can be used as the best egg white substitute in the bread business. Animal waste has been utilized by humans as building materials, textiles, and containers.

Figure 1.12. Gelatin.

(Source: https://upload.wikimedia.org/wikipedia/commons/thumb/7/7f/2019-10-10_22_15_43_Gelatin_from_a_single_opened_cup_of_Jell-O_strawberry_gelatin_snack_being_lifted_by_a_spoon_in_the_Franklin_Farm_section_of_Oak_Hill%2C_Fairfax_County%2C_Virginia.jpg/800px-thumbnail.jpg?20191011022258, Creative Commons License).

Gelatin (Figure 1.12) is another substance that can be made from animal skin and utilized in food. It can be used in a variety of dishes and is also a vital component of jellies. Collagen from skins can be used as an emulsifier in meat products because it can bind enormous amounts of fat, making it a useful addition. In surgical procedures, a substance made from removed collagen can act as a blood clotting stimulant. Animal intestines are cooked and then consumed as food in many different places. Brains, neurological systems, and spinal cords from animals are good sources of cholesterol and are also utilized to make cholecalciferol. As an emulsifier, cholesterol is utilized in the cosmetics business. Heparin, a substance formed from the liver, lungs, and lining of the small intestine, can be used as an anticoagulant to prolong the time it takes for blood to clot.

Because the growth of these industries depends largely on their ability to manage their waste, waste products from the poultry processing and egg-producing industries must be dealt with effectively. Fish waste is a great source of protein, lipids, and minerals. Shrimp and crab shells can be used to make chitosan, which has a wide range of uses in the cosmetic and pharmaceutical industries. Chitosan has numerous uses in the preservation of food, including direct application of chitosan film, direct chitosan addition to food or coatings onto food surfaces, and usage of chitosan-infused packaging materials. The dairy sector consistently generates significant amounts of proteinaceous waste, the majority of which is produced as waste water from milk tank cleaning and other dairy handling procedures. Dairy waste often contains a high concentration of dissolved organic components, including fat, lactose, minerals, and whey protein. Buttermilk, whey, skim milk, GR, and derived byproducts like caseins, caseinates, lactose, WP, etc. are just a few of the byproducts that the dairy industry generates. The primary byproduct of the dairy industry produced during the manufacturing of cheese is whey. Milk whey is high in proteins, soluble vitamins, dietary fiber, and lipids. The total amount of whey solids contains about 75% lactose.

The components of whey are separated using selective porous membranes and procedures called fractionation and isolation. Despite the fact that they can be employed as supplements in the agri-food sector, such as in sports drinks, 40% of whey is still wasted. Using lactose-consuming bacteria and aiming for bioconversion, one can achieve direct fermentation of whey, producing a variety of products. The co-metabolism of $C_6H_{12}O_6$ and galactose from lactose digested by *Pseudomonas hydrogenovora* can result in polyhydroxy butyrate. This bio-plastic is safe for the environment and has a significant impact on the medical industry. By creating nanoparticles, it can also make it easier for medications to be delivered to the body. Heteropolysaccharides like xanthan gum can be created from whey by utilizing *Xanthomonas campestris*. Due to their higher protein contents, lower energy, fat, and sodium contents, lack of microorganisms and lethal components, biocompatibility and generally recognized as safe (GRAS) status, availability and affordability, whey protein concentrate (WPC) and whey protein isolate (WPI) are widely employed in food applications. Caseins may be combined in various combinations and ratios with other ingredients to create composites (Ryder, Ali, Billakanti, & Carne, 2020). Caseins have also been used to create glues and adhesives for a variety of applications. GR has been used in food companies to make baked goods, and sweets, and as a flavor enhancer. GR (Ghee residue) is a fantastic source of phospholipids, which may then be extracted and added to ghee. Phospholipids, which are found in abundance in buttermilk concentrate, are employed in processed cheese spread to improve its organoleptic, and functional qualities. These byproducts can therefore be used in pharmaceuticals, food, health, and other industries.

1.9 ENZYMES AND THEIR USES

Enzymes, which are proteins, function as localized catalysts in certain chemical reactions when the conditions are favorable. These biomolecules are used extensively in a wide range of industries, including those related to food, personal care, medicine, textiles, chemicals, and fuels.

They are frequently chosen for chemical catalysis due to their higher level of substrate selectivity and other demanding operating criteria. Commercial enzymes are generally very expensive because of the high cost of raw materials and processing. By employing agro-food waste as the substrate for the production of enzymes, these rising costs can be reduced. Xylanases, amylases, proteases, invertases, peroxidases, polyphenol oxidases, pectinolytic enzymes, and tannases are a few examples. Two basic categories, submerged fermentation and solid-state fermentation are used to classify the production of enzymes. With submerged fermentation or SmF, the microorganisms are grown in excess of freely flowing media in either batch or semi-continuous operations.

While the soluble substrates are diffused in a liquid condition, the insoluble substrates are either submerged or suspended. Solid-state fermentation (SSF) includes the growth of microorganisms on moist, solid surfaces without the presence of a free liquid phase. Instead of bacteria, the development of fungi and yeast is possible using this method. SSF has an advantage over SmF (Submerged Fermentation) because it uses less energy, has a lower capital cost, and is more efficient. It becomes more appealing due to the lower capital expense associated with the downstream process and the simplicity of controlling bacterial contamination. SSF produces efficient environmental conditions that are similar to the natural environment of microorganisms for the development and production of enzymes. SSF cannot be used for large-scale enzyme synthesis because it is difficult to manage pH, aeration, oxygen transfer temperature, and moisture content.

Amylases are enzymes that break down glucose, maltotriose, and maltose, three examples of starchy carbohydrate units. Glucoamylase and amylase are the two main enzymes in the amylase family. By eliminating 1,4-D-glucosidic bond between sugar-related units along the lines of the amylose chain, amylase causes the hydrolysis of starch into $C_{12}H_{22}O_{11}$, sugar, and maltotriose, while GA hydrolyzes results in a reduction of amylose and amylopectin in glucose. These amylases are used in a wide variety of processes, including those that create bread, beer, paper, gin, and plant extracts. In the construction of amylases, carbon sources are frequently employed. Another method of producing amylase at a lower cost involves the utilization of extra agricultural wastes, such as food and kitchen waste (Figure 1.13).

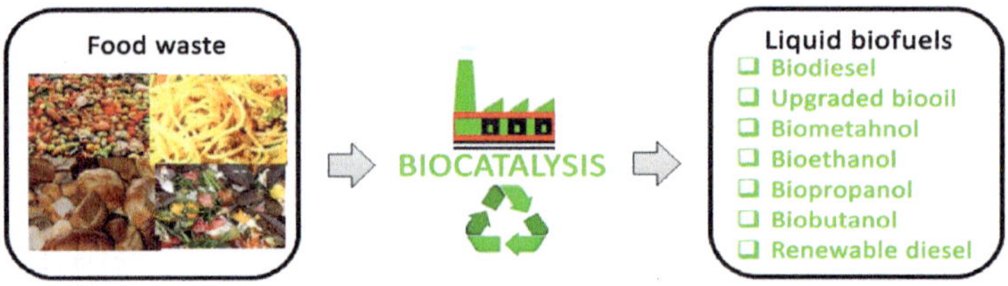

Figure 1.13. Kitchen waste and its uses.

(Source: Karmee SK. Moving towards the Application of Biocatalysis in Food Waste Biorefinery. Fermentation. 2023; 9(1):73. https://doi.org/10.3390/fermentation9010073, Creative Commons License).

Important hydrolytic enzymes known as proteases perform proteolysis by integrating hydrolysis into peptide bonds and binding amino acids to polypeptide chains. Due to their active involvement in cellular metabolism, application in food, medicine, detergents, dairy, and leather industries, and the purification of silver from discarded X-ray films, these enzymes are constantly in need of development. *Aspergillus awamori* produces protease utilizing bread crumbs as a substrate for the entire process (Melikoglu, Lin, & Webb, 2013b). The soybean cake was employed as a substrate to assist Penicillium species' protease assembly.The enzymes that break down xylan are known as xylanases, and they are a crucial part of plant polysaccharides. The complex polysaccharide known as xylan is made up of xylose residue and is linked together by beta-1,4-glycosidic linkages. D-glucuronic acid and D-arabinofuranoside may be connected to the Xylan. These are applicable to the food, feed for animals, biomedical, and bioethanol industries. The grape waste is what's left over after the fruit juice has been extracted. Due to its high phenolic content and low food content, this is not an acceptable animal feed. Pomace from grapevines is dumped, posing serious risks to the environment. Use of this value addition therefore seems to be advantageous. Grape pomace cannot be utilized for various fertilizers due to its high phenolic content, which inhibits seed germination.

Chapter

2 AGRICULTURE WASTE MANAGEMENT

Content

2.1 INTRODUCTION

According to estimates, China generates 56.2 million tonnes of agricultural waste (Aravani et al., 2022). The quantity and types of waste produced depend on the nation and how they are managed. Agricultural wastes are essentially the leftovers or residues from the harvesting, and processing of unprocessed agricultural products. Despite the fact that many agricultural wastes contain materials with value for humans, the majority of the time the cost of transporting, and processing the feedstock outweighs the economic value of these materials after processing. As a result, wastes are produced since it is not feasible economically to process or reuse rejected agricultural byproducts. For instance, large amounts of organic matter rich in nitrogen and nutrient-rich waste materials are produced along with foul odors, microbial contamination, and greenhouse gas emissions in areas that are specifically used for animal husbandry. In comparison to areas used for crop production, where minerals and organic matter are depleted due to the nutritional demands placed on the soil by growing crops, manure generation is consequently one of the most practical means of waste management for these areas. Important scientific and technological developments over the past few decades have led to a steady rise in agricultural productivity all over the world. However, this has also led to a constant rise in the volume of agro-industry by-products, agricultural crop residues, and livestock waste.

Despite the lack of reliable estimates, it is thought that agricultural waste makes a sizable contribution to the overall waste generation. The amount of agricultural waste generated was estimated to be close to 1000 million tonnes more than ten years ago (Han et al., 2003). It has been established that up to 80% of the total solid waste generated on a farm is made up of organic wastes. More than half of the fresh crop waste from 2012 was made up of organic wastes, which had an energy potential of 90 million tonnes of oil equivalent, greatly exceeding that of other waste like municipal waste, and tertiary forest remnants (Esposito et al., 2012). If the right measures are not taken, this situation is probably going to get worse. In the current era of industrialization and globalization, economic advancement is closely linked to the use of non-renewable resources, which has many negative impacts on the environment and biodiversity, jeopardizes the security of the food supply, and depletes our non-renewable energy reserves. Using specialized conversion methods, agricultural waste can be turned into commercially valuable items including fertilizers, electricity, and chemicals. This is essential to promoting a sustainable bioeconomy and, as a result, separating the use of non-renewable resources and economic development from human welfare and environmental sustainability. Even if a sizable portion of this agricultural residue is used as fuel, fodder, and for other home and industrial uses, there is still an annual surplus of 140 Mt that is not used, of which 92 Mt are burned. India produces more agricultural waste than Bangladesh, Indonesia, and Myanmar put together. In actuality, India burns more agricultural waste than the combined net amount of waste produced by the other nations in the region.

2.2 BIOMASS

Approximately 50% of all biomass is lignocellulosic. It contains cellulose, hemicellulose, and lignin and is the primary component of plant cell walls (Kim, Lee, & Kim, 2016). Agricultural residues (Figure 2.1) are those remnants from plants that are produced in agricultural fields. Field residues and process residues are two categories into which they might be divided.

Figure 2.1. Agricultural residues.

(Source: https://upload.wikimedia.org/wikipedia/commons/e/e4/Field_and_crop_residue%2C_south_of_Marlborough_-_geograph.org.uk_-_1229200.jpg?20110226094602, Creative Commons License).

Field residues are agricultural wastes that are still present in the field after crop harvesting is complete. Field residues serve as the starting material for a variety of processes intended to convert agricultural waste into commercially viable products like animal feed, organic manure, raw materials for various manufacturing industries, and even various biochemicals with clinical and industrial significance. Process residues are by-products that are nevertheless created by these processes. Various waste materials, including molasses, bagasse, seeds and seed pods, leaves, stems, straws, stalks, shells, and pulp, can be found in field wastes and process residues. India produces roughly 686 million tonnes of lignocellulosic waste, of which 234 million tonnes are agricultural wastes due to its agrarian culture(Kapoor et al., 2020). But a large amount of the agricultural leftovers produced in India is underutilized due to the current inefficiencies in agricultural waste management procedures. Proteins are a well-known component of our everyday diet and metabolic processes. They are also prospective suppliers of polymer materials. Aquaculture and livestock waste contain large amounts of protein and can be used to create new products with added value.

Figure 2.2. Aquaculture.

(Source: https://upload.wikimedia.org/wikipedia/commons/thumb/a/ae/Tiang_Sabah_Fish-aquaculture-01.jpg/800px-Tiang_Sabah_Fish-aquaculture-01.jpg?20160507091955, Creative Commons License).

2.3 AQUACULTURE

The growth of aquaculture (Figure 2.2) has led to higher feed for increased productivity in aquaculture. About 30% of the feed that is fed in a competent aquaculture plant is transformed into solid waste. The metabolic wastes produced in aquaculture are the most noticeable wastes. It might be suspended or in a soluble form. Because feeding rates rely on ambient temperature, a rise in temperature causes more species to eat. As a result, there is an increase in the amount of waste produced. Optimized water flow patterns in the production units are essential for waste management because they reduce the fragmentation of fish fecal matter and speed up the deposition of insoluble solid waste. Additionally, if there is a greater volume of non-fragmented waste, such as feces, it is simpler to capture them, which significantly lowers the concentration of dissolved organic waste in the system. Different sources produce the waste that results from livestock-related activities.

Figure 2.3. Animal farms.

(Source: https://upload.wikimedia.org/wikipedia/commons/thumb/7/7f/Farm_animals_in_spring_8a07.JPG/800px-Farm_animals_in_spring_8a07.JPG?20070411102149, Creative Commons License).

Animal farms (Figure 2.3) produce large amounts of animal feces and feed waste, whereas slaughterhouses produce organic materials and solid wastes like manure. The wastewater created by livestock activities includes urine, water produced during cage washing, water from bathing animals, and water from cleaning slaughterhouses. Air pollutants like H_2S and CH_4 are formed here. If these wastes are not properly managed, they harm the quality of the land and water and produce greenhouse gases. These pose a risk to the environment and to human health because they produce unpleasant odors and serve as breeding grounds for pathogens. Along with the growth in chicken product consumption, the poultry sector is also growing. The feathers that the birds leave behind are a major source of waste. The majority of this waste is made up of protein, particularly keratin, which is insoluble and difficult to break down naturally. Therefore, improved management is required to make use of these wastes.

2.4 WASTE MANAGEMENT

For the formation of an efficient waste management system, it is crucial to understand the characteristics of agricultural waste. Agriculture-related wastes come in a variety of sizes and qualities. Waste management is necessary for managing both solid and liquid wastes generated during agriculture processing and from animals. Understanding the properties of waste can aid in recycling, recovering usable parts, producing valuable goods like biofuels and biofertilizers, and general management. The type of product, techniques used to gather the raw ingredients, and processing techniques all affect the wastes generated during food processing. For instance, the waste produced by fruit and vegetable production is frequently acidic and rich in organic material. Wastes from activities like peeling and bleaching produce have higher levels of total dissolved organic solids. Vegetable parts contain water, carbohydrates, protein, and fat. Actions like washing add to the generation of wastewater. In addition, crop residues left over after harvest are another big waste source and a major concern, particularly in nations like India. They are made up of inorganic substances like cellulose, hemicellulose, lignin, nitrogenous material, and chlorophyll. To make quality control systems more effective, contaminants like solanine and dioxin that are present in both naturally occurring and man-made food wastes must be removed. Using fewer potentially dangerous items will result in less toxic waste being produced, making its handling easier. The killing, dehairing, removal of the skin, processing, and other cleanup procedures produce the majority of the waste in this. It comprises salts, processing chemicals, inorganic and organic materials, and blood. Typically, a beef slaughterhouse can create 50 pounds of blood, 50 pounds of dung, and 40 pounds of animal manure per cow (Jayathilakan et al, 2012). The amount and quality of waste produced are entirely dependent on how operations are managed. Water use, biological oxygen consumption, and solid waste production have all increased along with the growth of the poultry business. These sectors are not significantly impacted by seasonal changes, and as they grow, the amount of waste they produce is a major concern. Animal productivity as a whole and environmental changes can both be used to determine the features of animal waste. The amount of protein, fat, fiber, and non-fiber remaining in the feed after digestion is added together to determine the volatile matter that is present in the feces. By examining the contents of the lignin, cellulose, and hemicellulose, the fiber and non-fiber contents are determined. The properties of the waste are influenced by the composition and digestibility of the diet. The majority of the food that is excreted in the feces was not digested and was not broken down by bacteria. Unabsorbed excess nitrogen is eliminated through urine, and undigested proteins are a component of feces. Animal waste can be solid, liquid, or semi-solid. Therefore, the process of handling waste should be developed using trial and error techniques, depending on the type of waste. Newer techniques must be developed, and people must receive the appropriate training. Biomass is one of the currently available renewable resources that is widely available. It can be converted into other forms that are useful or utilized as an energy source. Future energy shortages are a major problem in the current environment. The energy crisis may be resolved by using biomass as a source of energy. Furthermore, the use of biomass is not just restricted to the production of energy; a number of industrial products may be produced using it. The necessity to create bio-based polymers has arisen as a result of rising plastic pollution (Figure 2.4) and poor recycling rates.

Figure 2.4. Plastic pollution.

(Source: https://upload.wikimedia.org/wikipedia/commons/thumb/b/be/Piles_of_plastic_waste_in_Thilafushi%2C_2012_%282%29.jpg/800px-Piles_of_plastic_waste_in_Thilafushi%2C_2012_%282%29.jpg?20210226155930, Creative Commons License).

For this, two main approaches have been used: first, the synthesis of biodegradable polymers from bio-based materials such as agricultural waste; and second, the use of non-biodegradable polymers modified to suit the needs of the situation. If the products can be successfully marketed and commercialized, both tactics will work. The physicochemical characteristics of the leftovers from alkali-treated raw cereal materials, including wheat straw and maize cob, were investigated. They contain a lot of dietary fibers and produce a lot of fibrous waste. The cereal materials and the insoluble fibrous residues, in contrast to the soluble fibrous waste, have important minerals, the capacity to scavenge 2,2-diphenyl-1-picrylhydrazyl radicals, good emulsification and emulsion stability, mineral binding capacity, and good water and oil holding capacity. Researchers have examined the potential of various agricultural wastes, such as tomato and grape pomace, green tea extracts, coconut shells, vegetable wastes, rice husks, fruit peels, and waste vegetables, in the production of biopolymers. The main challenge, however, is gathering the waste because the mechanisms for collection and sorting are insufficiently efficient to gather the waste. Due to the global heterogeneity of waste, typical production processes must change. However, commercial production of bio-based products derived from waste is required for global sustainability. When agricultural waste was analyzed in 2016 as a soil stabilizer, it became clear that India was not appropriately disposing of these pollutants (Vijayan & Parthiban, 2020).

About 85–90% of the silica in rice husk ash (RHA) is amorphous (Pode, 2016). The primary variables to be examined were vermicasts made from sugarcane bagasse, coir from coconuts, and tea waste. Therefore, there is potential for employing agricultural waste just to improve agricultural output. It could be utilized for a variety of purposes, including the creation of electricity, compost, animal feed, and more. Instead of being refined from the subterranean sources of fossil fuels like crude oil, biofuels are fuels that are formed from biomass. Through the alcoholic fermentation of glucose sourced from sustainable and regenerative sources like food waste and agricultural waste, bioethanol is formed. They have less of an adverse effect on the environment than traditional methods of ethanol production at an industrial scale. Aside from the fact that they use different substrates and have different carbon footprints, bioethanol and ethanol are chemically identical. One kind of biofuel is bioethanol. By 2023, ethanol is anticipated to account for around 67 percent of the worldwide growth in conventional biofuels, with biofuels like biodiesel and vegetable oil likely to account for the remaining 33 percent (Popp et al., 2016). The global production of bioethanol increased by 85 billion liters in 2011. Global bioethanol output and consumption are expected to reach 134.5 billion liters by the end of 2024 (Lemos & Mesquita, 2016). Agro-industrial wastes like rice straw, potato, sweet potato, and beet wastes, corn stalks, and sugarcane bagasse have all been investigated for the production of biofuel. Wastes with lignin and cellulose were prevalent in residues obtained from cereal crops, olive trees, tomatoes, and grapes. They may also be used as the primary raw material in large-scale ethanol production.

From lignocellulosic biomass, butanol is another product produced in addition to ethanol. It can also be blended with gasoline and, with only minor adjustments, used in place of gasoline in automobile engines. Lignocellulosic biomass is often produced from agricultural wastes like maize, fibers, and straw and is subjected to complicated pre-treatment processes that produce simple sugars that are then processed by microbial cultures that produce butanol. Rice straw, potato and sugar beet, sawdust, corn stalks, and sugarcane bagasse are meant to serve as a viable alternative to fuels derived from fossil fuels. Biofuels produced from agricultural waste or other analogous feedstocks recycle atmospheric carbon, which aids in slowing the effects of global warming. Agricultural residues are a better option than forest biomass because they are less expensive, simpler to get, and have a quick harvest time.

Figure 2.5. Deforestation.

(Source: https://upload.wikimedia.org/wikipedia/commons/thumb/d/d3/Deforestation_NZ_TasmanWestCoast_3_MWegmann.jpg/800px-Deforestation_NZ_TasmanWestCoast_3_MWegmann.jpg?20110613151236, Creative Commons License).

Deforestation (Figure 2.5) and the destruction of forest ecosystems are thus discouraged by the manufacture of biodiesel from agricultural waste. Growing suitable crops like non-edible oilseed crops just for the manufacture of biodiesel on farming waste lowers the need to dedicate valuable farmland for this purpose. A readily available and inexpensive energy source that is resilient to natural, social, and economic disturbances is essential for science, technology, and even social advancement. Biofuels derived from agricultural residues can be used to satisfy that requirement. Being an agrarian country, India has the potential to be a pioneer in the generation of renewable, sustainable biofuels from agricultural waste (Rao, Baral, Dey, & Mutnuri, 2010).

Another potential method that achieves the bioconversion of agricultural wastes high in organic material is the anaerobic digestion of agricultural wastes to produce methane gas. Methane is not produced during digestion in its pure form, but rather as a combination of gases known as biogas. This energy source is renewable and has several benefits. Along with the gas, it also creates extremely fertile manure that can be used on fields. Methane is produced as a result of the microbial fermentation-based digestion of agricultural waste in the absence of oxygen. Acid-forming bacteria first break down volatile substances to create organic acids, which serve as the food source for a group of microbes that manufacture methane. Methane is subsequently produced. Methane, carbon dioxide, nitrogen, and hydrogen gas with traces of H_2S make up biogas. The gas's heating value ranges from 18 to 25 MJ/m^3. This gas can be used to cook as well as generate power. The amount of meat produced worldwide has more than tripled in the last 50 years. Animal husbandry and livestock rearing activities have risen dramatically in most nations, including India, to meet the world's demand for animal proteins. However, the amount of waste produced by livestock activities has also considerably increased as a result of this. As a direct result of raising livestock, especially ruminant cattle, foul gasses like hydrogen sulfide, ammonia, and greenhouse gasses like methane are also emitted into the atmosphere. By solidifying the sludge and creating largely odorless sludge, anaerobic digestion makes the treatment and disposal of animal waste much simpler. As a result, it provides a different energy source.

India's tropical environment encourages the growth of weeds, pests, and harmful microbes in addition to crops. In order to prevent the rotting of their priceless harvest, farmers are therefore heavily dependent on chemical pesticides and herbicides. Farmers frequently wind up overusing and even misusing these compounds, despite the fact that their moderate use might assist boost agricultural productivity. Farmers frequently throw away the packaging and containers used to distribute these chemicals at retail establishments carelessly in open fields and bodies of water. The negative effects of pesticides and herbicides on the environment and its living ecosystems are generally underestimated by farmers. Unused pesticides and herbicides that are not properly stored can pose a serious risk to the environment because they may leak or seep out of their primary containment, contaminating the area. Chemical fertilizers are essential in contemporary agriculture (de Souza et al., 2020). Specifically, nitrogen, phosphorus, and potassium make up the majority of them. Fertilizers are also frequently misused by farmers in an effort to substantially increase their agricultural output, much like pesticides and herbicides are. The pace at which these fertilizer molecules are absorbed varies on a number of factors, including the physicochemical characteristics of the soil, the types of plants, and the fertilization method employed. A minute portion of the extra fertilizer is kept in the soil, but the majority eventually makes its way into other water bodies like ponds, lakes, streams, and rivers through surface run-off and irrigation canals. Air pollution results from a small portion of the extra fertilizer that evaporates or is denitrated. Therefore, natural fertilizers made from agricultural wastes of both plant and animal origin are preferable to traditional chemical fertilizers in the current environment. Chemical fertilizers can be used in conjunction with organic fertilizers or in place of chemical fertilizers entirely. Manure is a simple way to enhance the nutrients provided by chemical fertilizers. Poultry manure is high in phosphorus and has a good effect on agricultural output. Manure is added to soil to increase its nutrient retention or cation exchange capacity, water holding capacity, and overall physical condition and stability. This raises the soil's fertility.

2.5 CARBON SEQUESTRATION

Increasing the natural rate of soil carbon sequestration via soil amendment techniques is one of the areas of agricultural research that is being focused on because it can increase agricultural production. These methods have the power to reduce the amount of agricultural waste produced while also reducing greenhouse gas emissions. This is because the soil must hold carbon for a long period of time and be chemically protected to prevent it from being oxidized into carbon dioxide or reduced into methane. A long-term carbon sink with a very low rate of chemical transformation, partially burned products like black carbon and biochar have attracted growing interest for soil amendment applications over the years. Biochar is a porous, fine-grained substance that is high in carbon. It is made from the pyrolysis of biomass and contains various amounts of carbon, hydrogen, oxygen, nitrogen, sulfur, and ash. Biomass is heated in pyrolysis systems at temperatures between 400 and 600 °C while under anaerobic circumstances (Kim, Kim, Oh, Seong, & Park, 2022). This stops the substance from oxidizing, which prevents the emission of carbon dioxide. Instead, a portion of the biomass is vaporized, leaving behind a carbon-rich char that is known as biochar. When added to soil, biochar's porous nature helps increase the soil's surface area and ability to retain water. It actively affects and interacts with the soil microbiome, plant roots, and soil matrix, which helps retain nutrients and sets off a series of biogeochemical reactions that enrich the soil. There are numerous publications documenting soil quality improvements brought about by biochar addition, including better soil pH stability, an increase in earthworm populations, and a reduction in the need for fertilizer. Biochar has uses outside of agriculture as well. Water treatment, building, food technology, cosmetics, metallurgy, chemical engineering, and many other industries use biochar in a variety of ways.

Compost is a type of natural fertilizer created from organic residues like agricultural waste. They are created during the composting process. Composting is regarded as a risk-free way to manage waste. Microbes act on complicated degradable matter in an aerobic process to produce organic and inorganic byproducts. These repurposed waste products can be used as biofertilizers and to improve the soil's condition. It is also well known to increase agricultural production and yield in general. They are more resilient to toxicity, and drought. Compost encourages the cycling of nutrients and microbial activity in the soil, making it a viable alternative to damaging chemical fertilizers and pesticides. Composting agricultural waste on a large scale is a challenging project, but if done right, it will have a positive social and economic impact. However, the large-scale manufacture of organic fertilizers from agricultural waste is a difficult enterprise that necessitates a sizable initial technological, financial, and time investment. There have been publications on the main technical and financial issues that organic waste composting projects face. The absence of a consistent, secure, and stable market for the goods is one of the major problems. To create a self-sufficient supply chain, farmers can generate compost from their own agricultural waste and feed it back into their own land, so this is not a problem for small-scale, on-site fertilizer manufacturing facilities.

Figure 2.6. Composting.

(Source: https://upload.wikimedia.org/wikipedia/commons/thumb/c/c4/Sieving_the_compost_with_a_coarse_sieve_%2889165 52622%29.jpg/800px-Sieving_the_compost_with_a_coarse_sieve_%2889165 52622%29.jpg?20141029105842, Creative Commons License).

Composting, (Figure 2.6) a small-scale method of producing biofertilizers in a household setting, is another option that is practical and offers lucrative financial benefits, particularly for farmers who are struggling financially but have easy access to agricultural waste. Food waste, animal manure, and crop residue are high in organic content, making them perfect source materials for composting. Adding bio-compost to the soil significantly improved the agronomic qualities of wheat and rice. The majority of the agricultural waste materials used in the compost preparation included leaves, wheat straw, and rice straw. The biomass used included leaves, wheat and rice straw, vegetable produce, and compost was created. It was made up of solids, organic material, carbon, and all nitrogen. The compost created was abundant in organic material and carbon and gave the crops enough nitrogen and phosphorus. This promoted the growth of beneficial microorganisms and the local flora and fauna that are essential for healthy soil (Sharma & Garg, 2018).

The benefits of composting include increased productivity, disease resistance, and many other benefits. By employing the biodegradable waste that is burned, its usage should be stretched out. Agricultural wastes that are rich in nutrients, such as wheat and rice bran, maize cobs, husks, coffee bean waste, oil cakes, and uneaten fruit wastes, can be used by numerous businesses to create value-added goods. Potato peel powder was used as a substitute for some of the flour, and it had no effect on the fermentation or any other features. The CW ratio and temperature were the two key parameters that were examined. The leftover pieces of various fruits and vegetables are frequently discarded as waste since they are thought to be useless. However, numerous studies have demonstrated their value. For instance, it is known that pomegranate peels contain a number of bioactive substances like phenolic acids, flavonoids, and hydrolyzable tannins that have intriguing uses in natural food preservation. With the aid of substrates, solid-state fermentation can assist in boosting the antioxidant activity of diverse substrates. Pineapple waste can be employed as a substrate in SSF and produced antioxidants as well as anticancer drugs. It was shown that employing fermented pineapple waste boosted the level of protein, fiber, phenols, and antioxidants. As a result, it can be used as an antioxidant source. Environmental problems brought on by industrialization and urbanization include the excessive discharge of heavy metals. The fundamental issue with heavy metals is that, unlike organic contaminants, which can be broken down into non-toxic components, heavy metals cannot. Heavy metals' hazard to various life forms is another reason for concern. A variety of research has been conducted to address this issue, and it has been found that adsorption is a highly successful method for removing heavy metals in wastewater treatment facilities. But recently, agricultural waste has emerged as a practical, affordable alternative to the adsorption treatment of heavy metal effluents. Many by-products of agriculture, such as sugarcane bagasse, rice husk, sawdust, coconut husk, and oil palm shell, are considered low-cost agricultural waste by scientists for the removal of heavy metals. Agricultural waste was employed as an adsorbent to remove iron from drinking water (Sabir et al., 2021).

A number of studies have been conducted to evaluate the ability of agricultural wastes to act as adsorbents and remove heavy metals from water. They can be used to adsorb substances including nickel, chromium, lead, and cadmium as well as low-cost ion exchange materials. This will offer a practical method for water purification and aid in the control of agricultural waste. A variety of agricultural wastes and byproducts are created worldwide, and they have the potential to replace the current materials used as animal feed. Feed and related expenses account for between 50 and 60 percent of the cost of rearing animals. Therefore, creating a healthy feed that is also cost-effective is necessary to maintain the growing cattle. Despite efforts to identify alternatives, obtaining high-quality animal feed remains a problem in developing nations. The leftovers from harvesting are poor in protein, carbohydrate, and fat but a significant source of fiber. To satisfy the rising demand for livestock in the future, the traditional approach of feeding and raising livestock with fodder and pasture in addition to grain and protein concentrate is no longer effective. Therefore, feeding livestock with agricultural waste is an excellent alternative. Several agricultural by-products, including cottonseed meal, citrus peel, pineapple peel, rice bran, and soy sauce cake, were examined. The first difficulty encountered has to do with the economic and environmental effects of agricultural waste. Early forecasting tools are required since they can help users and policymakers. Life cycle assessment is a trustworthy process to be applied for measuring the impact of a product on environmental components. The problem in this, however, is the dearth of inventory information concerning the chain of agricultural residues. This is owing to the fact that the particular agriculture residue chain is either not evaluated at all or the data is accessible for a pilot-scale research. The second issue is the varied limits of the technology currently being used to turn agricultural wastes into valuable products. Anaerobic digestion, for instance, uses a variety of different agricultural wastes and is an efficient method for producing biogas and fertilizer, although it has some limitations. Energy crops frequently take the role of agricultural residues as the main feedstock in anaerobic digesters, which indirectly alters the pattern of land use. Overcoming the difficulties of creating molecules and materials from agricultural wastes is the third task. Only a small number of chemicals and polymers are currently bio-based. Finding alternatives is crucial due to the negative impacts of petroleum-based chemicals as well as the increasing scarcity of oil, which is utilized as a raw material, and other problems including the rise in plastic-based waste. The energy required, the nature and complexity of the product's chemical composition, the presence of pollutants, and general public opinion are the key challenges in the conversion of agricultural waste into biomaterial. However, the economic benefits added to creating electricity, animal feedstock, and other applications will be doubled if we are successful in making organic acids, enzymes, biodegradable plastics, and other items from agricultural waste.

The integration of agricultural waste and related businesses into the economy is the fourth problem. Due to resource variability, cross-chain valorization of agricultural waste is rather difficult. By-product streams are frequently bulky and expensive. To make valorizations possible, various production chains might be grouped into a single one. However, the primary products' economic values still play a significant role in company decision-making. There is slight consumer acceptance of products made from agricultural residues and a poor understanding of the value of valorization and associated potential in other sectors. The management of the material and accompanying information to address problems with nutrient imbalance, contamination, buildup, and residue conversion-related concerns is the fifth and final obstacle. Previous agricultural advancements in the EU contributed to the emergence of numerous environmental, socioeconomic, and technical concerns (Jones et al., 2017). While the supply of food and feed nutrients for export is frequently depleted, it is nevertheless in abundance in areas where animals are bred. For economic reasons, particularly as a result of agricultural and energy policy measures, the primary feedstock frequently gets substituted by certain crops rather than agricultural leftovers in areas with agricultural waste conversion technology like biogas plants (Figure 2.7).

Figure 2.7. Biogas plants.

(Source: https://upload.wikimedia.org/wikipedia/commons/thumb/d/d0/GAS-AG_Biogas_plant_in_Schwedt_%28074%29.jpg/800px-GASAG_Biogas_plant_in_Schwedt_%28074%29.jpg?20190310164920, Creative Commons License).

Because the conversion of agricultural leftovers depends on numerous industries, it's critical to raise awareness and encourage cross-sector cooperation. To ensure a circular economy where nearly no unrecoverable waste is produced, efficient resource utilization is essential. Wastes of many different types are produced by many industries. Biomass made from agricultural waste is essential for energy production in a circular economy. The utilization of waste biomass must be embraced from the producer to the consumer throughout the value chain process in a circular bioeconomic system. As a result, the product design and generated waste management systems must also be taken into account.

2.6 WASTE GENERATION

Waste production is escalating on a daily basis. By 2050, it is anticipated that the world's waste production would have roughly doubled (McKenzie & Williams, 2015). The developing nations will account for the majority of growth, and Asia is also where the majority of waste is produced. The largest contributor to any resource industry and the largest producer of waste in the economy is agriculture. The agriculture industry involves growing plants, animals, fungi, and other life forms for fiber, food, medicine, biofuel, and other goods that support and enhance human life. Waste management contributes to maintaining a healthy environment for farm animals in the right way. This may also lessen the need for fertilizers and other nutrients that are necessary for crop production. Through the local, regional, and global cycling of nutrients, sustainable agriculture is ensured by the appropriate management of agricultural waste biomass. Organic wastes, such as field waste, animal waste, agro-industrial waste, plant waste, and natural waste, make up the majority of agricultural waste. Rice, wheat, oats, maize, soybeans, barley, rapeseed, sugarcane, corn, sugar beets, and other crops are the main crops grown throughout the world, mainly in Europe, the United States, Canada, Brazil, Argentina, China, and India. Among the other agricultural crops, cereal crops were the main source of agricultural waste. Cereal crop waste includes stem and leaf materials, which made up 66% of the biomass in agricultural waste and were produced in 60% of the world's least developed and developing nations. About 23 Mt/per annum of agricultural dry waste were produced from cereals in the countries of the European Union; additionally, 368 Mt/per annum of wastes were produced in India in the form of straw. While China produced roughly 649 Mt of cereal waste each year (Tripathi, Hills, Singh, & Atkinson, 2019). Sugarcane is the second-largest producer of agricultural waste, while other agricultural biomass includes oil crops, roots, tubers, nuts, fruits, and vegetables.

About 517 Mt/per year of cereal crop residue went unutilized worldwide. The FAO estimates that by 2030, the amount of land used for agriculture would have increased by around 13%, or 120 M hectares, especially in developing nations (Fischer et al., 2010). Additionally, from 2005–2007 to 2050, there has to be a global increase in cereal production of around 0.9% (Jaggard, Qi, & Ober, 2010). As a result, it is possible to cultivate more crops on each plot of land, which could lead to an increase in agricultural waste. Similar to the cultivation of grains, there was an upward tendency for oil crops, which increased by 2.5 percent between 1999 and 2015(Jianyi, Yuanchao, Shenghui, Jiefeng, & Lilai, 2015). To meet their expanding needs, large economies like China and India have boosted their oil crop production. Rapeseed planting was accelerated in order to provide biofuel energy in European regions. The cereal straw is the product or seed that is left over after the crop has been harvested and processed. This dried stalk has a variety of uses, including compost for crop production, bedding for domestic animals, and fodder. This straw may occasionally be burned outside, in the field. However, due to air pollution, some nations have outlawed this practice. The compaction technologies that must be used to create straw pallets and wafers with high mass density entail a considerable cost. Rice husk is the most typical agricultural byproduct found all over the world. When processing rice, the husk is often separated from the grain since it is homogeneous in nature and has advantageous flow characteristics when compared to other types of biomass. The biofuel industry adapted several technologies, such as gasification, to achieve these features by using rice husk as a special fuel. There are some restrictions on the use of rice husk since it has a higher concentration of silica, which causes ash and slag problems in the boiler during the combustion stage.

The upper part of the corn is made up of the stalk, tassel, leaves, cob, husk, and silk. The remaining portion of it is used in the field, with just around 5% being used for animal food and bedding. Corn stover can be utilized as a fuel and turned into biofuels, as well as a fiber for paper industry/pulp manufacture. Compared to other cereals, sugarcane has better photosynthetic efficiency, which results in a relative increase in biomass production. 135 kg of sugar and 130 kg of dry bagasse are produced from each tonne of sugarcane, respectively (Masarin et al., 2011). In many sugar mills, dry bagasse, which has a heating value of 19.2 MJ/kg dry basis, is utilized as fuel to cook the cane and evaporate the water from the cane syrup. However, a substantial amount of bagasse is often disposed of using the inefficient burning method. All agricultural and food waste can be used for a variety of tasks, including making animal feed, building earthen homes, and most importantly producing electricity for some enterprises and industries, like sugar and rice mills. The wastes produced by various sources have an array of negative environmental effects on surface and ground waters, arable lands, and other habitats. By effectively using biomass waste residues, this negative effect can be reduced. The biomass wastes are typically burned in the power plant to produce energy, but the ashes and point source CO_2 can be combined to create carbonated materials that are mostly utilized in construction. These circular management systems can preserve landfill sites, strengthen the supply of building materials, and perhaps reduce CO_2 emissions and their associated negative effects by successfully utilizing biomass ash.

It is crucial to remember that not all kinds of biomass ashes have the capacity to undergo direct carbonation; nonetheless, numerous studies have revealed that some of them are easily carbonated due to the presence of minerals. In order to create composite materials, the non-reactive biomass ashes to CO_2 can be combined with the reactive biomass ashes in their raw texture. As a result, the production of carbonated-cemented composite materials, which are mostly utilized in construction, may be made with the highest care by mixing biomass ashes with raw wastes. The introduction of new bio-based products can be made by replacing traditional raw materials, which can ensure significant benefits to the environment, society, and economy of a country. In developing countries, particularly countries based on predominant agricultural activities, an enormous amount of waste was generated. The pressure on food production brought on by the rising trend in global population is the main cause of the massive waste creation, especially the agro-food waste. The majority of the produced agro-food waste is left unprocessed and not put to other uses. The untreated agro-food waste can be used for a number of things while still being environmentally acceptable. These wastes can be employed as animal feed, as a biofertilizer, as a building material for earthen dwellings, roads, etc., as well as for energy production as biofuel and heat generating in small-scale companies. By using fewer minerals, these wastes can also be used as a resource material to update contemporary tools with cutting-edge features. After generating heat, the created ash from industrial processes can be used for a variety of tasks, including the removal of hazardous metals from contaminated wetlands, the creation of glass materials, and the creation of biomedical, and optical tools. All of these methods are employed in the creation of low-cost, high-value tools that can lessen the effects of pollution on the environment and address waste management challenges by forging a viable substitute route.

Figure 2.8. Agricultural biomass wastes.

(Source: https://c1.wallpaperflare.com/preview/885/169/90/manure-wheelbarrow-wheelbarrows-waste.jpg, Creative Commons License).

2.7 AGRICULTURAL BIOMASS

The agricultural biomass wastes (Figure 2.8) or residues, which are often disposed of or burned as a source of feedstock materials, include crop stalks, leaves, roots, fruit peels, seeds, and nut shells. Utilizing biomass waste residues during the various crop processing stages of cultivation, harvest, and processing as well as consumer loss presents issues. Along with the waste produced during primary biomass processing, there are many stages of waste streams from secondary and tertiary sugarcane processing. It became clear that the amount of biomass from agricultural waste appeared to be roughly equivalent to the amount of byproduct produced during the growth, harvest, and processing of crops. Biomass waste was not properly handled in underdeveloped nations and was left exposed on cultivated ground to decay naturally. As a result, the majority of the biomass waste was left to accumulate, which had a harmful effect on the living things. The waste management process includes a number of steps, including waste material collection, transportation, processing, recycling, or disposal, and monitoring.

The collection, transportation, pre-treatment, processing, and final abatement of residues are the steps that make up the majority of the waste management system (Ayhan Demirbas, 2011). By reducing the amount of undesired materials entering or leaving the community and by raising knowledge of the reuse of materials within the community, waste management procedures primarily aim to provide a sanitary living standard. The waste management situation differs between developing and industrialized nations. Between these nations, the transfer of tested technology frequently seems insufficient. The characteristics and components of waste, seasonal climate variation, social structures and norms, the culture and behavior for handling and disposing of solid waste, their political structures, and the current awareness of the most predicted limited resource that is obvious to occur are just a few of the local factors that must be taken into account.

There are numerous creative ways to group waste according to traditional traits. The method of disposal is referred to as a cohesive strategy, which is necessary to handle a sizable amount of waste created from many sources during the production process. Over many years, the solid wastes were typically dumped in open terrain. This kind of waste disposal in landfills has grown in favor as the most economical method of waste management. Another waste management strategy that called for keeping waste in a container that was tightly sealed was landfilling. Typically, the storage containers have been built in a way that can stop different wastes from degrading during the landfilling process. By preventing the hazardous pollution brought on by the various sorts of waste, it contributes to maintaining the quality of the environment. Due to the lack of water and air, the enclosed containers help the usual waste and organic waste, such as papers, leaves, branches, and grasses, decompose more quickly. Aerobic bacteria remove the oxygen during the first phase of breakdown, which typically lasts fewer than seven days. The second phase, known as the anaerobic acid phase, is when the polymers are hydrolyzed into soluble sugars, amino acids, long-chain carboxylic acids, and glycerol using a range of hydrolytic and fermentative bacterial species. A variety of polymers, including cellulose, hemicellulose, proteins, and lipids, can be observed in the wastes. After a year from the start of the landfill process, the usage of methane appears to be beneficial if it has been commercially used. The production of home waste produces the main components of landfill gas produced from decomposed organic material. In anaerobic conditions, it typically occurs when naturally occurring microorganisms are present.

2.8 DRY AGRICULTURAL WASTE

Dry agricultural waste was discovered to be burned in developing nations for cooking and heat, primarily in suburban and rural regions (Cao, Zhang, & Zheng, 2006). It has been discovered that dry biomass can occasionally be transformed into liquid fuel when it is burned to produce electricity, which can also produce methane, hydrogen, and carbon monoxide. The fermentation technique was used mostly for wet forms of biomass that can provide both fuel and organic fertilizers, such as wet waste from sewage sludge, liquid food waste from the food sector, and so on. Direct conversion of biomass waste into liquid fuel, which may be used to power vehicles on the land, sea, and in the air, is simpler. Vermicomposting, also referred to as natural decomposition by microorganisms, is a process that occurs in the mesophilic environment. With different earthworm species present, various wastes go through the vermicomposting process. The waste substrate is degraded the earthworm intestines, which increases the breakdown of organic materials and enhances the physicochemical properties of the waste. The presence of microorganisms in the intestines causes the excrement to be biochemically degraded. In the end, the waste was turned into fine, homogenized manure that was rich in nutrients, a large number of microbes, humic compounds, growth hormones, and enzymes that promote the efficient growth of different kinds of plants. A circular bioeconomy is created as a result of the conversion of waste into valuable products, and this is crucial for assuring sustainable development.

Modern technology has enabled the extraction and use of carbon dioxide, with carbon dioxide being used as a feedstock. By converting CO_2, a number of materials, including polymers, fossil fuels, and building materials, have been produced. On the other hand, the effective application of CCU (carbon capture and utilization) technology will depend in part on the direct utilization of carbon monoxide and carbon dioxide emissions by applying cost-effective sorbents that can also be produced from waste. It's crucial to note that CO_2 is used to condition components like calcium silicate-based materials and concrete objects. The carbonation technique appears effective for solidifying cement-based forms and aids in stabilizing soil that has been polluted with lead, copper, and zinc heavy metals. By treating the waste through carbonation, regulated carbonation of biomass-derived wastes can play a vital role in risk reduction and the production of designed materials. The carbonation of biomass ash is crucial for maintaining its typical quality. If there is a rising trend of unregulated agricultural biomass dumping that can contaminate surface and ground waters, human health will be in danger.

The high agricultural biomass also contributes to eutrophication, GHG, NO, and N_2O emissions, which results in a faster rate of global warming than CO_2. Only a little more than two giga tonnes of crop residues are burned, which results in an 18% CO_2 emission worldwide along with large particles and black carbon. The components of biomass and how they are used for energy production determine the degree and kind of pollution from biomass. It appears that biomass has both beneficial and harmful effects on the environment. Positive environmental impact is described as the biomass renewability and carbon neutrality, while negative environmental impact is defined as the shift in land use, loss of carbon in soil along with nutrient reserve, depletion of biodiversity, and lack of water.

The use of coal and other natural resources identified in energy policies needed to be reduced (Reilly, 2012). These non-renewable resources can be replaced by renewable energy, and the sustainable use of agricultural biomass can reduce the consumption of natural resources. One of the best examples of this would be the United Kingdom, which employed alternative energy sources to meet its needs during the summer of 2019 and temporarily stopped using coal. Reusing biomass residues has undergone substantial development over the last 25 years. The development of biochar and other fuels was noted as a fruitful endeavor for producing power. Some agricultural biomasses contributed to the creation of renewable bioenergy. For instance, the synthesis of crop-based biohydrogen is aided by the use of maize, sweet sorghum, miscanthus, Kenaf, and eucalyptus from woody plants. However, the usage of biomass derived from microorganisms has increased, although plant-based residues have remained the top priority for fuel and feedstocks. Fossil fuels served as the foundation for economic growth around the globe for many years. All of these factors prompted people to consider a sustainable development plan that will guarantee increased crop production and energy production while using the fewest natural resources possible.

The agricultural areas were divided into tomato, pepper, watermelon, cucumber, and green bean categories based on crop output rates and levels of cultivation. These highly productive crop cultivation lands, which include both long-cycle and short-cycle agricultural crops, generate enormous amounts of agricultural waste biomass each year. The waste biomass residues, which are a type of plant waste made up of roots, stems, leaves, and fruits, were produced throughout the processing phase. In the province of Almeria, there was a lack of integrated management of greenhouse plant remnants, although there were significant prospects to enhance the recovery and reuse of agricultural waste in order to lessen its negative environmental effects. In the Spanish province of Almeria, several actions were made to implement sustainable waste management. Based on extensive study and detailed investigations on agricultural waste biomass produced from greenhouse crop lands, it was found that there was a significant potential for using these agricultural waste biomasses to effectively produce energy, biofuels, and bioproducts. When greenhouse agricultural fields in that province went through the processes of gasification, heating, recycling, and composting, the plant waste from those lands was regarded as an organic amendment. Moreover, a number of strategic and regulatory frameworks have been strengthened on a national and international level to introduce and implement the recovery of waste produced on crop land, a significant accomplishment to improve the usage of waste for various purposes. The waste from the two dominant crops, tomato and pepper, offers a chance to turn the waste into green fertilizers. The current global energy crisis is a barrier to long-term sustainability.

Dependence on fossil fuels, which make up 90% of energy use, causes climate change due to global warming and simultaneously depletes natural resources (Höök & Tang, 2013). Therefore, researchers are seeking innovative, sustainable, and renewable energy sources. In this sense, biofuels serve as an alternative energy source, lessen reliance on fossil fuels and greenhouse gas emissions, and promote energy security. In recent years, first-generation substrates such as wheat grain, sugarcane, corn, vegetable oil, or palm oil have been used to produce biofuels. However, as humans ingested these materials as food, the overuse of first-generation substrates indirectly increased food costs and contributed to the worldwide food crisis. Growing these food crops destroys natural ecosystems. As a result, the focus is now on second-generation substrates with high lignocellulose contents, including straw, wood scraps, or agricultural waste. Currently, the global agriculture sector produces 200 billion tonnes of lignocellulosic biomass per year. Thus, lignocellulosic food crops function as second-generation biofuels. Particularly, because it is a byproduct of food production, agricultural waste biomass is a good candidate for the development of biofuels.

However, increased food production had no negative consequences on the use of agricultural waste to produce biofuel. Agricultural wastes are inexpensive, widely accessible, and quickly biodegradable. Agri-residue is made up of both food-based components, such as sugarcane, maize, wheat, and beets, as well as non-food-based components, such as stalks, leaves, straw, corn cobs, rice husks, and orchard trimmings. Typically, agricultural residues are divided into three groups: food waste, livestock waste, and direct agricultural waste, which includes crop residues. Biomasses such as stover, stalks, cobs, bagasse, and other lignocellulosic wastes are examples of crop residues. The main sources of crop waste are the crops that are farmed all over the world: maize, wheat, rice, and sugarcane. Contrarily, livestock wastes are divided into three categories: solid manure, liquid or slurry manure, and wastewater, which includes disinfectants and processed farm water. Food waste has a significant potential for energy, making it an essential component in the formation of biofuels. Food waste is highly biodegradable due to microbial processes since it includes between 85 and 95 percent volatile solids and between 75 and 85 percent moisture (Mittal, 2006). Globally, the manufacturing of biofuel from agricultural remnants is expanding quickly due to economic and environmental concerns. Biodiesel, biohydrogen, biogas, bioethanol, bioethers, biomethanol, synthetic biofuels, biohydrogen, and bio-oils are all considered biofuels. These biofuels are sustainable and can be used as a replacement for fossil fuels, as a replacement for all energy sources altogether. Furthermore, biofuels have a number of benefits over conventional energy sources, including the fact that they are renewable, easily stored, reduce GHG emissions, foreign fuel dependence, pollution levels, waste management issues, improve energy security, stabilize energy crises, and support the growth of a new circular economy.

Figure 2.9. Fossil fuels.

(Source: Rehman A, Radulescu M, Cisma□ LM, Cisma□ C-M, Chandio AA, Simoni S. Renewable Energy, Urbanization, Fossil Fuel Consumption, and Economic Growth Dilemma in Romania: Examining the Short- and Long-Term Impact. Energies. 2022; 15(19):7180. https://doi.org/10.3390/en15197180, Creative Commons License).

2.9 FOSSIL FUELS

The vital sources of fossil fuels (Figure 2.9) are coal, oil, and natural gas. Due to their low cost and excellent efficiency, they currently provide around 90% of the main energy needed by the planet. Global CO_2 emissions from fossil fuels have been rising at a startlingly rapid rate since 1900. China is the main source of carbon emissions, accounting for up to 30% of all carbon emissions worldwide. The United States contributes 15% of all emissions, and India contributes 8% (Chhabra, Manjunath, Panigrahy, & Parihar, 2013). In the long run, CO_2 emissions increase global phenomena like climate change and global warming. Burning fossil fuels releases pollutants including particulate matter, NO_2, and SO_2 into the atmosphere. This strains water and land resources. In addition, the annual capital investment for the extraction and processing of fossil fuels has doubled from 2000 to 2013 (Smil, 2016). The world's growing population has increased our reliance on fossil fuels for energy production. This over reliance on fossil fuels for the production of electricity could lead to political and economic upheavals and uncertainty regarding the energy supply, which is known as the energy trilemma. While energy equality refers to the affordability of a certain energy, energy security deals with the steady supply of energy to the general public. In contrast, environmental sustainability concerns center on ensuring a low-carbon economy and lowering GHG emissions to slow global warming. Accordingly, switching from fossil fuels to renewable energy is a crucial driver toward achieving environmental sustainability. Renewable energy is energy that is continuously renewed by nature. It delivers only 20% of the current world energy consumption, which is significantly less than the anticipated global capacity. Solar, wind, hydroelectric, geothermal, tidal, and wave energy make up the main sectors of renewable energy. The most appealing alternative energy source among them is solar energy. Compared to fossil fuels, renewable energy provides a lot of advantages. Renewable energy cannot solve all environmental problems, even though it emits less greenhouse gas levels and lessens conventional air pollution. Installation of renewable energy projects impacts land use, wildlife habitat, and other factors. Understanding the environmental effects of renewable energy projects is therefore crucial for utility operations since it allows for better forecasting, planning, and execution to reduce or eliminate these issues.

As a result, governments and policymakers switched to a bio-based economy, which is more energy-efficient and cost-efficient. The biofuel business may offer an excellent way to lessen or perhaps eliminate these issues. Biofuels are typically produced from a wide range of feedstock materials. For the manufacturing of first-generation fuels, food crops are mostly employed as feedstock. Scientists have created biofuels from waste biomass and non-food crops to address this issue. Scientists have recently created third-generation biofuels to solve the issue of the diverse composition of feedstock and related conversion technologies of second-generation biofuels. Biomass from agricultural residues has the potential to be sources of renewable energy. The production of green energy from agricultural waste is seen as a key factor in the economic success of a nation. In addition to this, green energy, or the manufacturing of biofuels, lessens environmental effects like carbon emissions, global warming, and climate change. Because wastes are readily available at low prices, production costs for the produced biofuels are relatively low. Depending on the physical state of the biofuel, agricultural remnants can be converted into a variety of biofuels, including gaseous, liquid, solid, and other chemical forms. Gaseous biofuels include biogas, and biomethane, while liquid biofuels include biodiesel, bioethanol, biobutanol, bio-jet fuel, bio-oil, dimethyl ether/poly dimethyl ethers, green diesel, etc. Methane and carbon dioxide are both components of biogas. It is an economical, clean, and adaptable fuel that is good for the environment. A result of gasification or pyrolysis that contains H_2, CO_2, CO, and other elements is called synthesis gas, also known as syngas. It serves as an intermediate product in the processing of biomass. Syngas or biogas is converted into biomethanol, commonly referred to as wood alcohol, which is used in combustion engines (Demirbas, 2008). Bioethanol and biodiesel are both commonly used as transportation fuels worldwide. While bioethanol is an alcoholic biofuel mostly obtained from sugarcane, biodiesel and fatty acid methyl esters mixes are developed from either non-edible or edible oils.

2.10 LIGNIN

A wide range of agricultural wastes have different compositions. For instance, lignin, hemicellulose, and cellulose are the primary components of plant-based residues, whereas pectin, protein, carbohydrates, waxes, nitrogenous material, and inorganic minerals are present in smaller amounts. Contrarily, the primary component of cereals is starch, while the primary component of cow manure is protein. The building block of cellulose is cellobiose. It is a naturally occurring organic molecule that can be present in both crystalline and amorphous forms on Earth. Crystalline cellulose makes up the majority of cellulose, with a small amount of amorphous cellulose. Hydrogen bonds that bind the cellulose chains together are known as elementary microfibrils.

These fibrils are grouped together into bundles and joined by amorphous polymers, hemicelluloses, and pectin. Lignin covers the bundles of fibrils. Despite being resistant to both chemical and biological treatments, cellulose structures are nonetheless vulnerable to enzymatic breakdown for molecule hydrolysis in amorphous regions. Another vital component of plant cell walls and lignocellulosic biomass is hemicellulose. Hemicelluloses are heterogeneous branched polysaccharides with glucuronic acid and arabinose surrounding the xylan backbone. Either 1,3-glucan or 1,4-glucan linkages connect the hemicellulose backbone. The molecules, specifically xylans, galactans, mannans, and arabinogalactans, were primarily important for the construction of hemicellulose. Amounts of hemicellulose, which has a less complicated structure, ranged from 25 to 35 percent in agricultural residues (Chen, Lu, & Tsai, 2012). Hemicelluloses are typically found in amorphous form and are fully made up of sugars with either five or six-carbon molecules that have a molecular weight of less than 30000. As a result, they degrade more quickly than cellulose. Their composition and concentrations are influenced by the types of plants, tissues, and even molecules. A three-dimensional network of phenylpropanoid molecules made up the intricate chemical known as lignin. Phenylpropanoid units are typically joined by alkyl-alkyl, aryl-alkyl, and aryl-aryl ether linkages.

Lignin content is affected by plant type and age. Lignin offers structural stability, rigidity, and immunity from infections. Lignin offers resistance to both biological and chemical deterioration. Biofuels, aromatics, and macromolecules are the three broad categories used to describe lignin-derived value-added products. The main component of carbohydrates is starch. In most cases, it comes in the form of granules, and each one is made up of millions of amylopectin and amylose molecules. Typically, agricultural residue that has been air-dried has between 15 and 20 percent moisture, with the remainder being total solids (Tekin & Dalgıç, 2000). Volatile dry solids and trace amounts of inorganic materials, such as ash, make up the majority of the solid bulk. The energy is converted by this volatile dry solid. Ash contains several mineral elements, including calcium, magnesium, sodium, potassium, magnesium, silica, phosphorus, aluminum, and iron. Ash also contains nitrates, carbonates, sulfates, chlorides, oxalates, and both inorganic and organic amorphous compounds. Occasionally, depending on the circumstances, the remnants also include gaseous compounds of carbon, oxygen, hydrogen, nitrogen, and sulfur.

The main components of agricultural residue are carbon, hydrogen, and oxygen. On a dry weight basis, residue typically comprises around 35–50%

carbon and 40–45% oxygen. During energy generation, the final residual qualities determine the kind and amount of product. The fundamental understanding of residue aids in the modification of parameters during the energy conversion process. Against biological and chemical deterioration, lignin, hemicellulose, and cellulose form a barrier of protection. As a result, non-pre-treated residues hydrolyze slowly and yield is decreased. Pre-treatment is therefore required for effective biomass energy conversion. Pre-treatment generally improves pore size and surface area, decreases polymerization and crystallinity, increases responsiveness to treatment procedures, and results in higher yield. Pre-treatment should maximize the production of simple sugars, eliminate the lignin shell, prevent carbohydrate loss, reduce the production of inhibitors, and be cost-effective in its approach. Pre-treatment techniques typically fall under one of four categories: physical or mechanical, chemical, physicochemical, or biological. However, the choice of the pre-treatment technique relies on the kind of biomass. Size reduction through milling, grinding, or chipping is typically used as a physical treatment to increase the effectiveness of downstream processing. Furthermore, cellulose crystallinity is decreased through milling, grinding, or chipping. Vibratory milling outperforms other milling methods in terms of reducing substrate crystallinity and enhancing biomass digestibility. However, the amount of electricity needed for milling, grinding, or chipping varies on the kind of biomass and the particle size. In general, milling/grinding or chipping processes take a lot of energy.

Chapter

3

FARM WASTE MANAGEMENT

Content

3.1 INTRODUCTION

Disposing of waste in rural and urban areas has been a concern. Villagers are afflicted with a number of ailments as a result of the inadequate hygienic conditions, despite the open landscape. The main causes of rural pollution are improperly planned dwellings and drainage systems, as well as improper management of the disposal of plant, animal, domestic, and human waste. The correct collection of waste, a functional system for its transportation, and a safe and cost-effective method of disposal are fundamental criteria in rural regions for better sanitary conditions and the development of the general health of the rural community residing at the farm. Some attempts have been made to maintain and improve the environmental quality through the regulation of industrial pollution and the disposal of urban and rural wastes as a result of industrial development and the rapidly expanding population in cities and villages. Rural communities are largely disregarded when it comes to effective waste management and disposal (Henry, Yongsheng, & Jun, 2006). The types, amounts, and ways that wastes are used vary greatly from location to location. Every farm that generates waste must have a waste management system. The combination of waste generation, handling, storage, treatment, and disposal on almost every farm is distinctive. Rural industrial wastes are not important since there are no management or disposal risks associated with them. Although there are substantial amounts of waste from crops, animals, and people living there, they can be disposed of either by recycling or by discharge. Careful consideration must be given to this issue in order to establish new procedures and methods for the management and disposal of these wastes. As a result, the effective waste management and disposal system will also aid in the efficient use of waste materials for producing energy, serving as industrial raw materials, and serving as manure for agricultural areas.

3.2 WASTE

Waste can also be defined as materials that are damaged, defective, or unnecessary that are produced during the manufacturing process, materials that are discarded from agriculture and forestry, and waste that is available from places where people and animals live that are either allowed to escape without being used or are used on the site. Waste materials are consistently viewed as materials that are more cost-effective to throw away than to store for use. Waste is no longer seen as undesirable and should not be put outdoors. Therefore, with enormous economic and environmental benefits, all material that is in the wrong place can be separated, moved, repurposed, and reused. All wastes produced by human and animal activities that are typically thrown away are collectively referred to as solid wastes. Discarded waste materials are frequently recoverable due to their inherent qualities and may be regarded as a resource in another situation. Waste materials are all those materials that are formed both during production and during consumption or utilization. A material only ceases to be valuable in the production process and at the location where it is created for this reason. These wastes are without a doubt highly valuable in terms of manure, energy, and industrial raw materials. For the generation of methane and manure, the operators of biogas plants can employ resources such as soil from households.

Figure 3.1. Solid wastes.

(Source: https://upload.wikimedia.org/wikipedia/commons/thumb/1/1a/Lack_of_solid_waste_management_leads_to_blockages_of_the_drainage_%2869153295 21%29.jpg/800px-Lack_of_solid_waste_management_leads_to_blockages_of_the_drainage_%2869153295 21%29.jpg?20141031022058, Creative Commons License).

Solid wastes (Figure 3.1) may result in the deterioration of air quality, surface and groundwater contamination, and land pollution (Biswas et al,

2010). Leachate is produced when water seeps through the wastes and can eventually mix with groundwater. The air quality has declined close to disposal sites as a result of wind-scattered dust and waste. Additionally, as waste decomposes, foul gases are released that pose a risk to human health. The effects of inappropriate waste management on the environment have been researched globally. It is now widely recognized that improper solid waste collection and disposal procedures contribute to the spread of several disease vectors and water-borne diseases. By using appropriate approaches for managing urban and rural solid waste, the majority of environmental consequences can be reduced. Government owned and managed services now only collect an average of 50% of the waste that is produced; nevertheless, for cities to be regarded as reasonably clean, at least 75% of these waste quantities need to be collected. As urban populations rise and per capita waste generation rates rise, which are anticipated to rise by one to three percent annually, so will the need for services. Open dumping is used for disposal, usually on flood plains and into ponds, which has a negative impact on the ecosystem. Unfortunately, there is no adequate system in place in any city for managing solid waste, from collection to disposal. Uncollected waste poses an important threat to public health by obstructing drains, creating stagnant pools, and serving as a breeding ground for mosquitoes and flies, which increases the risk of cholera and malaria. An important amount of the collected waste ends up in dumping grounds, open pits, ponds, rivers, and agricultural land due to a lack of proper disposal sites. Environmental deterioration gets worse over time as the population, urbanization, industrialization, and overall economic growth of the nation increase.

In recent decades, there has been a migration from rural to urban areas. The main causes of this migration are the absence of alternative work options, the slow development of the agricultural sector, low crop yields, and environmental deterioration brought on by water logging/salinity, deforestation, and deserts. In turn, the substantial inflow of rural residents has contributed to the overstrained condition of metropolitan infrastructure and services. Housing, drinkable water, transportation, power, gas, drainage, and sewage are just a few of the essential urban resources and services that have rapidly declined in quality and availability. The municipal institutions lack the technical capability and resources necessary to meet the demands of the growing urban population. Urban poor communities frequently participate in a number of self-help projects, such as recycling and solid waste management. Nearly all of the glass, metal, plastic, and paper is collected and recycled. As a result, waste recycling is crucial to the disadvantaged communities found in urban centers. Only organic waste is still present on the streets. Composting could be done with this, but neither rural nor urban regions have embraced this business to its full potential. All solid, semi-solid, and liquid materials that the owner no longer considers to have sufficient value to keep in urban or rural regions are considered wastes. For the design and operation of functional aspects related to solid waste management, knowledge about the kind, sources, composition, and rate of formation of solid waste is essential. In both urban and rural communities, sources of solid waste are typically connected to zoning and land use. These organic compounds are made up of things like paper, cardboard, textiles, plastic, rubber, and leather, as well as farm wastes. Glass bottles, tin cans, aluminum, ferrous and non-ferrous metals, batteries, oil, paints, and other materials are examples of inorganic components. Paper, cardboard, plastic, glass, and other packing materials are prevalent in commercial waste. This includes waste from public and private institutions, as well as schools and colleges. Similar to residential and commercial waste, this type of waste may also include hazardous and non-hazardous wastes, such as chemicals from laboratories. In this kind of waste, paper, cardboard, and packing materials make up a majority of the waste. These wastes, which are categorized as industrial wastes, are produced through a variety of processes in both small and large-scale companies. These are industry-specific and rather varied in character. Industrial wastes contain both hazardous and non-hazardous components (Pappu, Saxena, & Asolekar, 2007).

Waste produced during the construction, repair, destruction, and rehabilitation of buildings and other structures is referred to as construction and demolition waste. These comprise materials such as plaster, cement bags, stones, concrete, mud, timber, and plumbing and electrical parts. These are the wastes generated during the use and maintenance of municipal facilities, such as street sweeping. It comprises farm debris from public parks, play areas, and roadsides, as well as dead animals and tree clippings. Agricultural wastes also include animal excrement and waste from dairy farms. Sludge from water and wastewater treatment facilities is one of the solid and semisolid wastes. These wastes' properties vary depending on the kind of treatment facility. These wastes demand unique handling and disposal procedures. Waste generation is necessary because undesired materials must be thrown away for disposal. It is a constant activity that is difficult to manage. Municipal solid waste is produced in enormous amounts throughout the megacities. The amount of municipal solid waste produced varies depending on the way of life of the population. The population is growing gradually, although only seven nations account for 90% of the annual rise in population, including Pakistan, India, Bangladesh, and Indonesia. The rapid population growth of megacities in developing nations has been a result of uneven planning. Megacities like Islamabad and Karachi are growing in population by 500,000 people annually. According to all estimates, managing MSW will be an important concern for years to come in all developing countries.

It is important to highlight that there is an important disparity between the amount of solid waste generated in Pakistan and the amount that reaches final disposal sites. Before it is finally disposed of, a lot more waste from developing nations, like Pakistan, is recovered, typically by scavengers. Any data pertaining to the quantification of wastes must be analyzed in light of the aforementioned in order to have any significance. Statistics can also be inaccurate when predicting how much waste is generated by households and then finally arriving at disposal systems. Pakistan's predicament is made worse by the absence of weighing facilities at the majority of disposal sites as well as a tradition of waste sampling and examination. Municipalities are beginning to understand the value of weighing, and most recently weighing facilities have been built at disposal sites in Lahore. The kinds and amounts of waste that are generated and recovered vary by location and, to some extent, by season. Regions with more traditional lifestyles tend to produce relatively little waste, and segregation and reclamation procedures are more common. The amount of solid waste is rising along with the population and level of wealth. This association makes sense and is consistent with global experience. The amount of solid waste produced in any nation varies according to the style of living of its citizens. The wealth of the population contributing to the waste affects the composition of municipal waste. To determine the best choice for disposal and recovery, it is critical to understand the composition of waste, both at the source and during disposal. For instance, among other things, the amount of waste produced and the percentage of organic waste define whether composting is feasible or not. In rural areas, where many waste materials are utilized, the organic content of solid waste is lower.

3.3 DATA ON WASTE IS PERTINENT

For the purpose of choosing appropriate treatment and disposal techniques, the data/information is necessary. MSW (Municipal solid waste) contains an important amount of biodegradable wastes, such as food wastes and farm wastes, is excellent for composting (Thapa et al, 2022). There has been a lot of research done on MSW composition. The exact composition is determined by the area, time of year, lifestyle, type of land used, etc. Food wastes, paper, cardboard, plastics, rubber, textiles, leather, farm wastes, wood, glass, tin, aluminum, and other metals, as well as silt/dirt and building and demolition wastes, are important components of MSW produced in the cities. Municipal solid wastes can exhibit important seasonal fluctuations. Seasonal wastes include mango peelings, melon peelings, waste from sugarcane, and other fruit and vegetable wastes. These seasonal wastes are present in large quantities, which drastically changes the composition of MSW. Waste composition varies from locality to locality as well. There are frequent similarities among residents of a certain area in terms of their earnings, preferences, and spending habits. Paper and packaging are frequently found in high-income locales' waste, but food wastes are typically the main component in low-income localities.

In places where these operations are taking place, a sizable amount of waste is made up of construction and demolition waste. The composition of wastes from commercial locations is influenced by the types of operations. Paper and packaging are typically the main components in the area around offices and institutions, whereas food wastes are more common in the area around vegetable and fruit markets. Similar to how waste from slaughterhouses frequently contains bones, blood, and animal body parts, waste from dairy farms will be substantial in animal feed and feces. An integrated waste management plan is necessary for effective waste management. There are numerous techniques, and the selection is largely influenced by the waste compositions. There is an important amount of plastic included in solid waste, which is extremely concerning. Every stage of the production process releases plastic waste, and every plastic product that is used up becomes waste afterward. Environmental issues are brought on by plastic waste, both in terms of quantity and quality. This is true because they come in high volumes but light weights. The recycling industry is responsible for handling this waste. Littering, incinerator emissions of dioxins and hydrogen chlorides, and pollution from chemicals are a few of the environmental problems caused by plastic waste. Wildlife is directly endangered by plastic waste. By doing away with plastic bags, compost quality will improve and there will be less waste that needs to be disposed of.

3.4 MUNICIPAL SOLID WASTE

Municipalities are primarily responsible for collecting solid waste. In high-income parts of big cities, waste collection efficiency ranges from 90% to 0% in low-income rural areas (Turan, Çoruh, Akdemir, & Ergun, 2009). The amount of waste collected is significantly lower in many other regions, especially in the poorer regions where burning at the primary point of disposal, natural biodegradation and dispersion, and local self-help for disposal to informal dumping sites are frequently the only options for disposing of solid waste. The municipality is primarily responsible for managing municipal solid waste. The government has historically been in charge of collecting waste from homes in middle to high-income districts while municipalities are in charge of sweeping the streets. Public waste collection is typically ineffective, and the main reason for this is a lack of funding. As a result, there is a trend toward outsourcing a sizable portion of waste collection and street sweeping services to private firms. Municipalities are not currently engaging in any sort of recycling effort under the current system. The majority of the time, individuals keep the principal recyclables including paper, plastic, glass, and metals and then sell them to street vendors or waste dealers to be recycled. Agricultural, residential, and rural industry wastes are all included in farm waste. Animals and plants are the sources of agricultural waste. Most of these wastes are currently either not used at all or used inefficiently. The ways in which waste is handled, treated, and disposed of have the potential to negatively impact the quality of the air, water, and land as well as to annoy individuals who live nearby. A waste management system is necessary for any farm that produces waste products, and practically every farm has a unique combination of waste generation, handling, storage, treatment, and disposal, either through recycling or discharge. Unsanitary living practices, as well as inadequate management of domestic wastes, are some of the prevalent aspects of rural zones. These include stagnant pools of water, carelessly handled waste, and free junk polluting the streets. Rural industrial wastes are not substantial, and as a result, they do not constitute a serious risk to management and disposal. However, sanitary conditions can be preserved and some financial gains can be obtained if correct disposal techniques are used.

3.5 DOMESTIC SOLID WASTE

Regardless of the size of the city, domestic solid waste is currently not handled in a proper manner during collection, transportation, and disposal or dumping; as a result, the environmental and sanitary conditions have gotten worse every year, and people suffer from living in such conditions. The range of issues with solid waste management is very broad and entails taking into account every aspect directly or indirectly connected to solid waste and its management. The rate of urbanization, the pattern and density of urban areas, physical planning and control of development, the physical composition and density of waste, temperature and precipitation, the activity of scavengers for the separation of recyclables, and the ability, sufficiency, and limitations of the respective municipalities to manage the solid waste, i.e., storage, collection, transportation, and disposal, are some of these factors. The discipline of waste management is concerned with regulating the production, storage, collection, transfer, and transport of wastes as well as their processing and disposal in a way that adheres to the best practices in public health, economics, engineering, conservation, aesthetics, and other environmental factors. All administrative, financial, legal, planning, and engineering tasks involved in finding solutions to the waste issues that the community's residents have imposed upon it are included in waste management. Due to the quantity and variety of waste produced today, financing restrictions, the impact of technology, and the lack of social awareness in the rural community, challenges with rural waste management (Figure 3.2) are complex.

Figure 3.2. Rural waste management.

(Source: https://upload.wikimedia.org/wikipedia/commons/thumb/7/79/Bonasse_waste.JPG/800px-Bonasse_waste.JPG?20150316131622, Creative Commons License).

There isn't a single solution for all waste management issues. Due to the availability of several waste management techniques, economic analysis can be used to identify the least expensive option (Gnansounou & Dauriat, 2010). The type and volume of waste generated and utilized, as well as the treatment costs determine the least expensive waste management technique. Many organizations have attempted to turn wastes into useful goods or, less frequently, into disposable effluents that are safe. The majority of wastes that are collected are difficult or expensive to use. The adoption of a new waste utilization technique depends on how much it can boost financial performance through an improvement in the cost-revenue relationship. Waste utilization has generally become a topic that requires careful consideration. Products that use waste and how those wastes are used are influenced by market forces. Utilizing waste is therefore more than just a technical feasibility issue. The optimal level of reuse is not produced by the current industrial organization system. If utilization plays a part in the waste management plan, then the market will decide whether or not new technology will be successfully adopted. Therefore, analyses of the economic viability of waste utilization processes should be included in diverse initiatives to develop new uses of waste.

There are many benefits to considering waste as a potential source, including its abundance, free of cost, adaptability, labor-intensive utilization, and legalized recycling. Solid waste is largely produced by both producers and consumers. The volume of wastes increases with population size and wealth. It is unclear how quickly waste will increase as the population and standard of living rise. However, it is a fact that the disposal issue is getting worse, not only because the urban population is expanding quickly but also because the nature of goods and services is constantly changing. All researchers and interested parties, including doctors, engineers, social scientists, and others, always place a high focus on health. When designing urban and rural areas, town planners and environmental engineers are the most concerned. Although there is a lot of open space in the villages, the main issue is that residents have a variety of diseases because of the unsanitary conditions. The villages need to be made aware of the necessity for effective management and disposal of domestic, animal, and human waste, which is to blame for environmental degradation and health issues. The finest management and disposal strategies can be discussed in public lectures, seminars, radio talks, telefilms, and other forms of media. Town planners and environmental engineers always place a priority on health. Planning for villages can also be done in this manner (Wheeler, 2013). Despite having an open landscape, the people are plagued by a number of ailments as a result of inadequate sanitary conditions. Poor drainage, stagnant pools, waste dumps all over the place, unprotected water sources, etc. are characteristics of rural communities.

Urinating outdoors creates an offensive smell and poses a health risk since it can spread diseases. Even in affluent neighborhoods, buses frequently stop and passengers exit to use the restroom. There are very few bathroom facilities, and those that do exist are poorly constructed, poorly kept public restrooms and urinals. All cities, both large and small, and their villages, lack adequate public facilities. As a result, damage from solid wastes is likely to outweigh that from other types of pollution. Since the beginning of human civilization, both humans and animals have exploited the earth's resources to sustain life and get rid of waste on the planet. The early population was often small and dispersed, and there was a lot of area accessible for the assimilation of wastes, therefore disposing of human and other wastes did not provide a serious difficulty. The ability of the fundamental components of a natural ecosystem, namely air, water, and land, to self-clean, was unaffected.

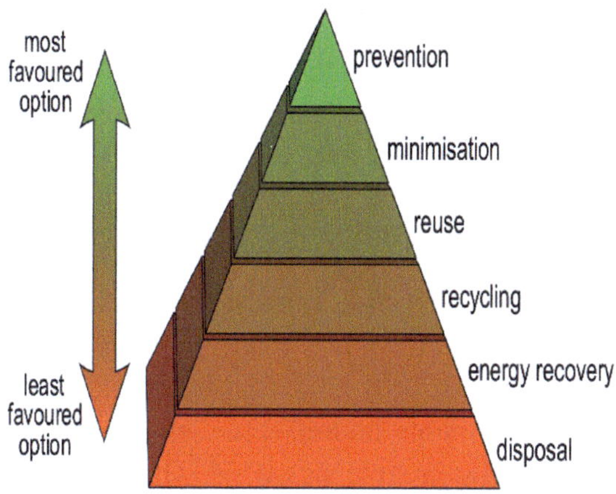

Figure 3.3. Waste management.

(Source: https://upload.wikimedia.org/wikipedia/commons/1/18/Waste_hierarchy.svg, Creative Commons License).

Waste management (Figure 3.3) and disposal may be traced back to when people first started to live in villages, and communities and the buildup of wastes became a threat to human health. The plague pandemic and other diseases have been brought on by the lack of any plan for the management of rural and urban waste. Public officials didn't start taking public health measures seriously until the nineteenth century when they realized that in order to manage disease vectors, food wastes needed to be collected and disposed of in a sanitary manner. Solid wastes in both urban and rural areas are inevitable in any culture. In some villages, stagnant ponds and other similar areas of land are reclaimed using rural waste. However, if surface soil layers are not sufficiently covered, this can also result in unsanitary surroundings. Such open dumping not only severely pollutes the air and ground, but also renders that piece of land unusable for any purpose until the pollutants disintegrate. In the past, people dumped their waste in rivers and oceans or buried it. Economic factors are currently the main factors taken into account for waste utilization. These include accessibility, chemical content, useful physical qualities, collection and transport, waste material separation and environmental protection pressures, energy, end-product quality, and social attitudes in use. Therefore, energy generation, raw materials for industries, and manure worth may be viewed as the economic value of agricultural waste. High energy use is a reflection of a high standard of living, which is linked to GDP.

3.6 FUEL NEEDS

The majority of today's energy needs are met by fossil fuels like coal, wood, petroleum, and natural gas. Coal reserves make up about 90% of the estimated fossil fuel reserves (McGlade & Ekins, 2015). In villages, firewood and locally accessible livestock dung provide around 80% of the overall energy. However, if correctly managed, the biomass present in the communities is adequate to supply the immediate requirement. Through the use of contemporary technology and effective management practices, some of the heat potential can be greatly increased. These wastes can be used to produce energy by anaerobic fermentation to produce biogas, which can then be effectively employed. Other rural wastes, like night soil, manure from chickens, and animal urine, can also be utilized to make biogas and fertilizer. Numerous crop products that are either underutilized or unutilized can be used as raw materials in a number of businesses. These byproducts include sugarcane bagasse, straw, husk and bran of rice, straw of wheat, paddy, barley, and pearl millet. Either these items are composted or they are disposed of as manure. For a variety of industrial uses, straw and husk from paddy, wheat, and sugarcane are also employed. Yeasts can use manure as a substrate for fermentation, and it could be possible to extract alcohol from the process. Due to its high protein content, returning feces to animals is becoming more and more common. The by-product of animal corpses is not effectively used. After being properly utilized, the bones can occasionally be used as cow feed or to make bone ash, bone charcoal, dicalcium phosphate, etc. Other by-products, with the exception of skin and bones, currently have little to no utility despite the fact that there is a great deal of potential for using them. Currently, the main sources of manure are cattle dung and different crop wastes. The fertility and structure of the soil are enhanced by farmyard manure. Dead animal byproducts can be utilized as manure. While dried and powdered meat can be utilized as a source of nitrogen manure since it has a high percentage of nitrogen, bones can be used as phosphate manure.

The night soil that is left out in the open either ferments, dries, or partially ferments and partially dries. The dust and dried debris combined to spread airborne infections throughout the neighborhood. The water supplies of the settlements were also combined with this. The wastes are excellent sources for boosting soil fertility and have a very high manure value. Therefore, residential waste, unutilized crop residues, animal excrement, and their by-products, all of which are considered waste are in extremely abundant supply in rural areas. A methodical approach to management and usage utilizing contemporary tools would not only keep rural regions clean but will also supply enough energy, manure, and raw materials for numerous initiatives. In the modern sense, nothing is wasted, and even waste that has the potential to give us inexpensive energy is often ignored. The vast majority of the materials found in the rural wastes, which are generated in enormous amounts every day all over the world, are always of an organic character. Currently, the majority of rural waste is dumped in so-called sanitary landfills, which serve no greater purpose. There are undoubtedly other ways to put farm waste to use. Both large and small cities have designated sites where waste can be dumped. Unexpectedly, although being quite unsanitary in nature, they are called sanitary landfills. These landfills not only have an extremely offensive odor, but they are also overrun with cockroaches, rodents, flies, and other potentially dangerous species. The decomposition process is always connected with the emission of gases, an important component of which is methane, which is a positive element of landfills. The primary component of all biogas is methane, which has a high fuel value. Organic material in the buried waste is progressively broken down by microbes to release free methane. Because of the slow release, it is difficult and expensive to tap this resource for use as fuel. However, this energy is still being harnessed and utilized in many locations, notably for the creation of electricity. Water addition to landfills will hasten the biodegradation process and concomitant methane production. According to estimates, each cubic meter of landfill gas might create enough energy to power 1,000 rupees worth of electricity. There is a hotel complex in the US that utilizes electricity generated from landfills (Jingura & Matengaifa, 2009). Smaller towns and villages will therefore be in a favorable position. The amount of organic matter in the waste produced in these places is higher, which will result in more gas being produced per unit volume of waste, making the process more appealing.

3.7 RURAL WASTE MANAGEMENT

For the creation of workable waste management systems, characteristics of rural waste must be addressed. These are crucial in deciding the reuse of waste, liquids, transportation of raw materials, recovery of certain waste components, irrigation, byproduct development, energy, and fertilizer value, among other issues. Similar opportunities for reuse exist for solid wastes, such as a source of energy, animal feed, raw material for industries, and a soil conditioner. Rural waste is a heterogeneous mass that comes from farming, animal husbandry, residential areas, and rural businesses. The quantity, physical features, and chemical composition of the waste will determine the optimal design of the treatment and disposal systems. More operational flexibility, reserve capacity, improved environmental pollution controls, and less expensive environmental processing equipment must all be incorporated into the design of the waste management system. The starting point for design is provided by the characterization of the typical content and qualities of wastes. As a result, planning for potential changes in waste characteristics needs to be taken into account.

On farms and in villages, there are varying degrees of waste, organic waste, and dead plants. The elements of waste also differ from location to location based on the economic situation and eating patterns. The majority of food waste is fed to animals, therefore there is only a very tiny amount of non-edible vegetable items high in cellulose that are thrown away. The majority of the vegetation finds its way into the waste in, warm regions with heavy rainfall. However, in many villages, there isn't enough of this waste to even combine with the night soil to make a suitable biodegradable pile. The weight of a material per unit volume is referred to as bulk density. The value can change depending on the presence of water. Most solid wastes have bulk densities that are importantly lower than the final densities of their constituent parts. Solid wastes have low starting densities and poor compaction, which reduce carrying capacity and add to the high cost of collection and transportation (Ferronato et al, 2021). To minimize the volume and enhance the density of waste, as well as the amount of area needed for the final deposit, compaction is frequently necessary. The specific particle size density and pile porosity needed by a certain combustion process are provided by the densification, gasification, and burning of biomass. The moisture content of the waste is one of the variables influencing compaction characteristics. The fluid's resistance to shearing when it is in motion is measured by its viscosity. Wastes that are liquid or semiliquid become less viscous as their temperature rises. The lowest temperature at which the material may be easily pumped should be taken into account.

3.8 PLANT WASTE

The main sources of plant waste include the roots, stems, leaves, and pods of field crops as well as the products of underutilized plants. The fruit industry as well as plantation crops may produce an important amount of waste. From crop to crop and variety to variety, there are different types and amounts of plant waste available. The woody part of the plants is typically burned, while the non-woody part is either used as animal feed or piled into manure in the traditional manner. It is not a waste substance, but the underuse of it results in important waste. While preserving sanitary conditions may not benefit from efficient management and disposal of plant wastes, it may benefit farmers. Unused or underused commercially uncultivated plants can produce a wide range of beneficial goods, including fiber, gum, resin, rubber, tan, color, and lac. However, waste does not apply to the description of commercially uncultivated plants, their derivatives, or crop residue utilized as feed. Given that cellulose is a key component of the majority of crop leftovers, the paper and pulp industries may be the best places for them to be used. Three crucial raw materials for this industry are paddy, wheat straw, and sugarcane bagasse. Through a comprehensive research and development program, it is necessary to investigate the use of other crop leftovers for papermaking. For the manufacturing of pulp, cellulose with a low lignin content is the ideal raw material. Crop leftovers can be used to make a variety of papers, including fine paper, blotting paper, filter paper, packing board, straw board, greeting card, album paper, and cover paper, depending on their individual features.

Villagers use waste materials like paddy straw, cotton stalks, and wheat straw to build dwellings. A cementing material or binder can be made from the husk, which makes up 20 to 25 percent of the entire rice processed. To make roofing sheets that the underprivileged can utilize, sugarcane bagasse, rice straw, wheat straw, and jute sticks are employed. Even developed nations like the USA are expressing interest in making paper homes for low-income groups. The key steps in the production of paper roofing sheets include pulping, board-making, corrugation and drying, impregnation, fire retardant treatment, and finishing. The chemical energy in the fuel is transformed to heat energy through the exothermic process of direct burning. Agricultural wastes from a variety of crops, including cotton waste, groundnut shells, sugarcane bagasse, corn cobs, and straw, are burned directly for fuel. Less than 10% of these wastes used in home chulhas are thermally efficient. Burning agricultural waste in furnaces, such as rice husk, groundnut shell, and paddy straw, can boost the efficiency of heat usage. In the process of compressing and reducing intra-particle spaces, densification aids in both gasification and direct burning of plant wastes.

3.9 COMPACTION

The densification process comprises roll compaction, pelleting, briquetting, tableting, and bailing (Bajwa et al, 2018). Solid waste contains a comparatively high cellulose content, making it acceptable for use in solid fuel-fired furnaces when appropriately treated. Size reduction using a hammer-mill shredder is the first step in the densification process. Although the material has been shrunk, the size distribution is not consistent. The heavy portion is eliminated after size reduction by passing the solid waste material through an air classifier system. The majority of the fine material is eliminated during the screening process after the light material is released. Following screening, the light fraction is palletized. There are three different forms of agro-waste equipment, including manually operated, bullock-operated, and power operated. Individuals and families in rural or village settings can collect waste agricultural material that is dispersed in these settings, convert it to briquettes, and use the briquettes as home fuel by using a manually powered compaction machine. Pyrolysis is the process of using heat to break down complex molecules into smaller, simpler molecules in an inert atmosphere. Pyrolysis and incomplete combustion are frequently used interchangeably in the context of fuel conversion. The majority of organic materials are thermally unstable. They can be divided into gaseous, liquid, and solid parts after heating in an oxygen-free environment via a combination of thermal cracking and condensation processes. This process is importantly more exothermic than combustion, as opposed to the endothermic pyrolytic process.

3.10 CLEAN FUEL

A clean fuel can be produced via the pyrolysis process to replace fossil fuel. It changes the heterogeneous wastes into a homogenous form, which is preferable. The char can be utilized as a filter medium, landfill, or soil conditioner and has fuel value. The liquids may also have chemical value in addition to their utility as fuel. Compared to natural gas, the gas has a comparatively low heating value. It is difficult to store and not cost-effective to transport across long distances. Because of the high amount of carbon monoxide it contains, handling it calls for extra safety measures. However, the pyrolytic process requires close observation. The efficacy of neutralization is increased by briquetting or pelletizing the agricultural base, although the smoke may still be generated. The heating properties of the biomass may be comparable to those of domestic coal when it is carbonized or charred under controlled conditions and then made into char briquettes with or without a binder. On a modest scale, the charring can be done in an empty oil drum. In around 8 hours, 15 to 25 kg of biomass can be burned. After mixing with any available binder, this burned material can be manually pelletized or extruded using a hand extruder. Except for the cost of manual labor, production costs are quite low. For rural locations, a community-level charging machine with a batch capacity of 1 to 3 tonnes has been designed (Jouhara et al., 2017). The unit can be constructed permanently using regular bricks with a side lining of fire bricks to retain heat. To feed biomass and remove burnt material, a number of fixed trays with corresponding charging doors are present. About 400 kg of biomass is required to supply the heat of pyrolysis for two times the amount of biomass in the charring unit. The system's key benefit is that no electricity is used. Unskilled laborers can complete the task since it is simple. Given the expense of collecting raw materials, it is preferable that the plant with the necessary capacity be situated close to the location where raw materials are available.

The pyrolytic process, which produces coal from home and industrial waste, uses a variety of equipment. To gather plant waste for the manufacturing of pyrolytic coal, a waste collector is a manually operated, pull type rake. The rake can be constructed from bamboo or mild steel.

If the leaves are dispersed over the area, it takes an individual an hour to gather one quintal of dried leaves. A waste cleaner is an inexpensive piece of manually operated machinery that can separate household waste from combustible items like paper, leaves, fruit and vegetable peels, and textile remnants. Waste is ground in order to make it denser and reduce the amount of waste that needs to be transported. At the point of collection, the grinder can reduce plant waste to less than 25 mm in size. As a result, a 100 kg gunny bag may hold around 20 kg of plant waste rather than 3.5 to 5 kg if it is unground. The equipment being utilized weighs no more than 15 kg and has an hourly quintal capacity. The household waste pyrolyzer is constructed from a regular metal drum that can contain 10 kg of dry plant material. The waste material is ignited from the bottom holes while the other holes are covered with lids in this drum. Due to the chimney that is retained over the drum, fire rises. The first layer's holes close when the fire reaches them, and the second layer's holes open when it does. In this manner, fire spreads from the second to the third layer while being put out below the openings in the second layer due to an oxygen shortage. The waste is thereby transformed into pyrolytic coal powder.

Processes for producing different types of gases using gasification were primarily created for peat and coal. However, compared to coal, biomass contains far more oxygen and hydrogen, which lowers the quantity of steam required for gasification. Furthermore, it often has a low sulfur concentration. On the other hand, as biomass frequently contains a large quantity of water, if it is not properly stored after drying, it will draw an equilibrium amount of water. In addition to increasing the process' complexity, this results in energy losses during drying. Because the biomass is frequently in an awkward shape, substantial feed preparation, such as grinding and pelleting, may be necessary unless the gasifier is particularly adapted. The gas engine is cooled to nearly ambient temperatures and thoroughly cleaned to have it ready for use. The cooler cleaner assembly properly pre-processes the gas needed for the engine to have a long and trouble-free life. Utilizing water to scrub surfaces results in cooling. When a steady supply of cooling water is not available, tar is removed by condensation, soot and ash are removed by centrifugal methods, and air cooling is used. Prior to entering the engine, pure gas and air are properly diluted in combination. The char is transformed into CO and H_2 when this hot gas rises through the hot material (Wang, Weller, Jones, & Hanna, 2008). The biomass is pyrolyzed by the hot gases at the following level, and then the biomass is dried by the gases at the top of the bed. Low temperature gas is produced by heat exchange. This gas contains a lot of tars, oils, and water, which can interfere with proper combustion. The oils are passed over the heated char and break into gases, yielding relatively little oil. Because of this, downdraft gasifiers are especially well suited for operating internal combustion engines. An advantage of a fluidized-bed gasifier is that it can treat a wide range of biomass sizes and has a high heat transfer rate due to its extremely high circulation rate. Due to their short contact times, they are less effective than downdraft gasifiers for consuming char or cracking oils and tars. There is a propensity for the light char fraction and light biomass fraction to prematurely separate from the bed. However, fluidized-bed gasification technology is still in its infancy. High throughputs and the ability to treat microscopic particles are two potential benefits of suspended fuel gasifiers. To get to a stage where it is feasible, suspended gasification technology needs to be properly examined and research and development work must be done.

3.11 SUGAR FOR FUEL

It is very likely that the production of fermentable sugars from cellulose hydrolysis will increase global ethanol and food supply. Vegetable sugars are fermented to make ethanol. For every 1.5 kg of sugar, one liter of ethanol is produced. Molasses can serve as the raw material for ethanol production. At room temperature, concentrated acid easily depolymerizes and dissolves cellulose, resulting in a combination of sugars. Economically isolating and recovering the corrosive acid from sugar is the main issue with concentrated acid operations.

The cellulose is hydrolyzed with diluted acid at a high temperature in the dilute acid procedure. The main issue with this method is that the acid, when heated, also catalyzes the breakdown of sugar into byproducts. Without recycling, dilute acid procedures can only convert cellulose to hexose by 55 percent. More than 80% of the conversion of cellulose to hexose can be achieved using recycling. The diluted acids in cellulose are likewise particularly corrosive at high temperatures. Cellulose is hydrolyzed to hexose through enzymatic mechanisms. Several bacteria create the celluloses. There is no degradation of the sugar generated as a result of the mild enzymatic reactions. The main drawbacks are the high cost of the enzyme used per unit of glucose production and the requirement for expensive, energy-intensive pretreatments in order to achieve high yields. Although it can be utilized in compression ignition engines with particular engine modifications, ethanol is best suited for spark ignition engines.

Bacteria that work completely without oxygen provide anaerobic treatment. These bacteria consume intricate organic molecules, which they break down to provide energy and serve as the basis for their own structural building blocks. The digestive process consists of two stages: liquefaction and gasification. Large organic molecules like carbohydrates, proteins, and lipids are broken down into smaller molecules like simple organic acids and alcohols in a process called liquefaction. The simple acids are typically gasified simultaneously, producing methane and carbon dioxide. Methane and carbon dioxide gases are created when plant waste is treated anaerobically (Derbal et al, 2009). As a combustible gas, methane can be burned to provide energy. Bacterial breakdown of the organic solid waste is done in order to apply plant wastes as manure. This is referred to as humus. Composting refers to the full procedure that involves both the separation and bacterial conversion of organic solid waste. The opportunity to recover and reuse some of the nutrients and organic material in solid waste is provided by composting. Composting goals include stabilizing putrescible organic matter, preserving crop nutrients and organic matter to the greatest extent possible, and producing a consistent dry product suitable for use as a soil conditioner, garden supplement, or for land disposal. The nature of the starting material, the circumstances of the composting operation, and the degree of decomposition all affect the chemical and physical properties of humus in different ways.

3.12 COMPOSTING

Composting occurs in biological systems that include oxygen. In general, aerobic composting is characterized by a high temperature, the absence of unpleasant odors, and a faster rate of decomposition than anaerobic composting. Anaerobic composting is distinguished from aerobic composting by its low temperature, generation of odor-producing intermediate products, and generally slower rate of progress. The fundamental benefit of anaerobic composting is that it may be done with less supervision and, as a result, can be shut off from the environment. The ambient temperature is often the required intermediate temperature for mesophilic composting. Composting that is thermophilic is done between 45 and 65 °C. In actuality, both ranges are present in most operations. The types of compost systems that are classified as open or windrow are those where the entire process is done in the open. Typically, the material is stacked in long windrows. On the other hand, with a mechanical system, the digester is an enclosed unit where the majority of the first composting action occurs. Environmental elements that affect biological activity, such as moisture content, temperature, pH level, nutrient availability, and oxygen concentration, regulate the rate of composting. The amount of water present in the composing pile has an important impact on how quickly materials decompose. Large amounts of heat are produced during decomposition, and if there is not enough water in the pile, it will tend to dry up and the rate of decomposition will nearly completely stop. Composting has a theoretically 100% moisture content. However, the actual moisture content depends on the net aeration capability as well as the composition of the material being composted. For the majority of fibrous vegetative waste, the maximum permitted moisture content is from 75 to 85 percent, while for waste of green vegetation like wet waste, it ranges from 50 to 60 percent. If the maximum allowable moisture content for a certain waste is too low, the moisture content of the waste can be increased by adding an absorbent. So, if straw is added to vegetable waste, the allowed maximum wetness increases proportionally to the amount of straw supplied. If the waste's moisture content in its raw state is very high, an absorbent must also be added. Bacterial activity can occur at moisture contents as high as 12 to 15 percent.

An important environmental component influencing biological activity is temperature. Cryophilic, mesophilic, and thermophilic temperature ranges are the three sub-ranges that life typically occurs in. Thermophilic and Mesophilic temperatures are the ranges in which modern composting processes are intended to function. Because there are so many different types of organisms involved in the process, the ideal temperature range for composting is rather wide, ranging from around 35°C to about 55°C. Except at the beginning of the process when the material is at ambient temperature, a constant temperature does not exist throughout the mass of composting material unless a carefully regulated digester is utilized. For most bacteria, the ideal pH range is between 6 and 7.5, and for fungi, it is between 5.5 and 8.0. Early in the mesophilic digestion process, the pH drops from around 6.5 to 5.5. It then rises quickly during the thermophilic stages, reaching just over 8.0, before dropping to 7.5 during the cooling and maturing stages. Due to the requirement for nitrogen conservation, little can be done in a practical operation to change the pH level currently present in the pile. Lime will have the effect of raising the pH above what is desired and removing the nitrogen as ammonia if it is used to elevate the pH in the early stages. Natural aeration through the porous material usually supplies the pile with oxygen. The pile benefits from periodic rotation, which promotes aeration, increases thermophilic activity, and keeps the moisture content and temperature distribution constant. It is thought that turning composting windrow piles once a day or once every several days achieves effective aeration. For fine material, oxygen can reach a depth of 15 to 20 cm in a windrow pile.

Figure 3.4. Windrow composting.

(Source: https://upload.wikimedia.org/wikipedia/commons/thumb/7/78/Windrow_buidling8_%2814728274800%29.jpg/450px-Windrow_buidling8_%2814728274800%29.jpg?20141027175737, Creative Commons License).

In windrow composting, (Figure 3.4) moisture can be controlled in a variety of ways, from removing extra moisture to adding water when a pile is too dry (Mohee & Mudhoo, 2005). With variable degrees of success, moisture can be managed by adding straw, dry soil, sawdust, etc. Care must be taken to prevent an excessive increase in the C/N ratio, though. Frequent rotating, which reduces moisture content without changing the C/N ratio, is advised as a treatment for excessive moisture. Compost that is left outside in an area with a lot of rain may get very damp. In this situation, it would be wise to take preventive action by covering the piles with a roof. Severe rainfall only allows 3 to 6 mm of moisture to permeate composting material. Using a hose to sprinkle water yields comparable outcomes. Water can be added more often while a pile is being turned to increase its moisture level. Water is absorbed by the exposed material. In clear weather, the exposed portions of open heaps dry out; in this situation, surface sprinkling is advised to encourage uniform decomposition. Unlined pits are more likely to have their walls fall apart, giving the pit an uneven shape. The material in the pit can be turned as frequently as required to maintain the appropriate high temperature and aerobic conditions.

The walls and bottom of the pit offer some insulation against heat and moisture loss when the pits are used, exposing a lesser portion of the stack surface to the air. For aeration and drainage of liquid seepage from the pile, the pit may have a chimney, trenches, or a porous bottom. It is advised to compost in pits that are about 0.9 meters deep using a method that maintains high temperatures and aerobic conditions for the first few days and anaerobic conditions for the next four to five months. At least 15 to 22 cm of waste is placed on top of the material when it is heaped in the pits. The aerobic organisms can produce a high temperature for the first few days because there is enough oxygen in the initial stack. Despite the fact that anaerobic conditions develop after the first few days, it appears that the high temperature is maintained for around two weeks because of the insulation capabilities of the stack. For roughly three months, the material is allowed to compost in the pit without being turned, primarily in anaerobic circumstances. Another layer of material may be added on top when the pile settles to preserve the appropriate depth. When the pile enters anaerobic condition, earth or another covering material can be added to it to help absorb escaping nitrogen, hold in moisture, and stop flies from reproducing. The top 10 to 15 cm of the material in this configuration typically does not decompose efficiently. Less is known about whether sufficiently high, homogenous temperatures have been created to eradicate the disease. Pits cannot be utilized effectively when the groundwater table is high; if they are used, drainage from the bottom of the pit is preferred, and provisions should be made to stop surface runoff from entering the pits. The pit approach has the advantage of conserving labor, but at the cost of more insects, odors, and a less assured eradication of pathogens.

3.13 C/N RATIO

The preservation of nutrients and fertilizer value of organic waste is one of the most crucial goals of composting. It's crucial to conserve nitrogen. Nitrogen conservation is more challenging than that of phosphorus, potash, and other micronutrients (Aulakh & Malhi, 2005). Leaching of nitrogen may occur, but the main loss of nitrogen occurs when ammonia or other volatile nitrogenous gases escape from the compost material and enter the atmosphere. The C/N ratio, pH, moisture content, aeration, temperature, form of nitrogen molecules at the start of composting, and the adaptive or nitrogen-holding ability of the composting materials all affect nitrogen loss as ammonia during aerobic composting. A ratio of roughly 30 or higher between the available carbon and nitrogen allows for the least amount of nitrogen loss. Numerous researchers have established the ideal C/N ratio to avoid nitrogen loss under various circumstances. When conditions are favorable, the nitrogen fixation may take up a small quantity of extra nitrogen. Nitrogen is lost in the first phase, stabilized and conserved in the second, and recycled, conserved, and occasionally accumulated in the third. This demonstrates that the majority of the nitrogen in waste may be conserved by composting activities. As pH increases to roughly 7.0, ammonia exists as ammonium hydroxide more and more quickly.

Materials with a high starting pH can be predicted to lose more nitrogen than those with lower levels of ash. A portion of the ash should always be kept out of the compost piles and added later after composting. Although less important than the C/N ratio and the pH, the moisture content of the compost has an impact on nitrogen conservation. When the moisture content is low, there is more ammonia leakage. The water acts as a diluent and solvent for the ammonia, lowering its vapor pressure and preventing it from volatilizing. The conservation of nitrogen will help when the moisture level is between 50 and 70 percent, which is also suitable for other aspects of composting. Nitrogen conversation is negatively impacted by aeration. The more the substance is spread out and exposed to the atmosphere, the easier it will be for any ammonia to escape. The losses in turning won't be important when other variables support nitrogen conservation. Since high temperatures are essential for aerobic composting and the eradication of pathogens, they also promote the volatilization and release of ammonia. Reducing the temperature after the first two turns or after the first six to eight days of active decomposition will not importantly improve nitrogen conservation because the majority of ammonia is lost during the early stages of active decomposition. The effect on nitrogen loss will be minimal if the other elements conducive to nitrogen conservation are well maintained and the temperature is below 65° to 70°C. Nitrogen conservation may be impacted by the initial form in which nitrogen is contained in the substance. Even though the C/N ratio is sufficient for nitrogen conservation, if the raw materials contain substantial levels of ammonia, part of it may be volatilized and lost before the organisms have had enough time to use and stabilize it.

Given that an important amount of nitrogen is lost during the first few days of composting, this can be an important factor. If the composting material has an excessively high initial moisture content, the extra liquid drains away, which can also cause nitrogen loss by leaching in wet weather. The amount of soluble nitrogen in the compost and the amount of rainfall both affect the loss of leaching. By preventing water from entering the compost stack, leaching can be reduced.

Chapter

4 PRODUCT RECOVERY

Content

4.1 INTRODUCTION

The global food and beverage business is one of the biggest and is vital to all economies. This reflects its contribution to the necessities of every living thing. As a result of the rapid rise in global population and the corresponding rise in affluence, the demand for food has significantly increased during the past 50 years. In developed countries, where money is no object but time is scarce, rising food consumption has been followed by a rise in food processing, with a focus on the creation of energy-demanding, ready-to-eat canned, frozen, dried, and fresh foods. Such processing is now being done globally as globalization spreads across all industries, and many final products are subsequently shipped to the right markets. There are a number of economic and technological factors that contribute to the large amount of food processing waste that is created (Mirabella, Castellani, & Sala, 2014). Traditional food preparation techniques produce very tiny amounts of waste that would previously have been disposed of as compost, as animal feed, or by municipal waste disposal. The majority of food processing systems were created at least 20 to 30 years ago, when waste disposal, particularly in the industries that process vegetables, grains, and fruits, was not a concern. In the past, there was no motivation to find other ways to deal with wastes because the value gained by processing a fraction of a raw food source to produce a high-priced product outweighed the expenses of disposal. In fact, such a strategy would significantly increase corporate risk from a strategic perspective. Because of this, creating technologies and methods for utilizing wastes was not given as much attention; instead, waste creation was still essential for the advancement of food processing systems. From a business management standpoint, the challenge of utilizing food processing waste may be taken into consideration. Waste food processing co-products are not easily employed by the processors due to the specificity of a certain raw material and its processing in connection to a specific product. Exploiting the waste would need some diversification, perhaps involving the creation of new products for existing or emerging markets. The players in the food market would face a high level of risk, which is not appealing given that the sector is already mature and that the majority of the products are commodities. It is hardly unexpected that processors prefer that outside parties remove their wastes from their facilities. The difficulties of dealing with wastes are made worse by the quick degradation of biological materials as a result of autolytic, chemical, and microbiological spoilage, which reduces their potential for use as food and the value that comes with it.

4.2 WASTE

Waste generated during the production of food is primarily biodegradable and results from the processing of biological resources. Biowaste is defined as a waste capable of undergoing anaerobic or aerobic decomposition. The waste can come from a diversity of sources, primarily from plants and animals, but it can also come from fungi and bacteria. The various phases of each operation will generate a lot of waste; these have also been genetically defined.

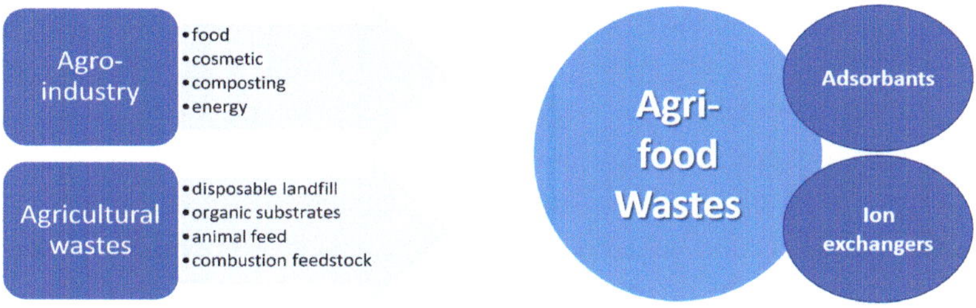

Figure 4.1. Biological wastes.

(Source: Matei E, Râpă M, Predescu AM, Țurcanu AA, Vidu R, Predescu C, Bobirica C, Bobirica L, Orbeci C. Valorization of Agri-Food Wastes as Sustainable Eco-Materials for Wastewater Treatment: Current State and New Perspectives. Materials. 2021; 14(16):4581. https://doi.org/10.3390/ma14164581.jpg, Creative Commons License).

Biological wastes (Figure 4.1) can range from full, unusable materials to fractions and combinations created by physical, thermal, chemical, and biological processing of the original raw material. They are extremely complex since they are derived from extremely intricate living species. Plant wastes range from pulps and residues left over after the extraction of oil, starch, sugar, and juice to vegetable and cereal remnants like bran and extracted barley grain, and more. Wastes from the manufacturing of dairy products as well as wastes from animal-derived sources include blood, bone, and brain tissues. A lot of effluent is frequently produced. Producing foods made of both plant and animal ingredients frequently results in the creation of more complex mixed wastes. Because of the effects of the many processing activities on the intricate raw materials, food processing wastes have a heterogeneous composition.

4.3 BIODEGRADABLE WASTE

The biodegradable components in food wastes and effluents are abundant and have high biological oxygen demand and chemical oxygen demand contents (Porwal, Mane, & Velhal, 2015). They pose a threat to the environment if they are not handled and treated because of the poisonous chemicals and methane that are produced during their uncontrolled decay. Numerous waste management solutions have been established at the national and international levels in an effort to lessen the impact of waste. For instance, better systems management may address waste that occurs from errors in process management, such as excess substrates, malfunctions, or human error. Contrarily, process waste may need to be addressed through process design improvement and may necessitate the purchase of new equipment. When properly purified, some wastes, like water, can be recycled, aiding in the creation of closed-loop manufacturing facilities. The production of innovative products can be used to take advantage of co-products that cannot be recycled. This tactic might not be appealing because it is probably a high-risk one for the food processor. However, this choice is becoming more and more relevant as a result of growing governmental pressures and customer demand. Whey from the dairy industry and starch from potato processing are two examples of very large co-product streams that have been effectively used to produce food components. This, however, took between 10 and 20 years to happen, reflecting the length of time it takes for new technologies and markets to develop.

4.4 RESIDUES

Small quantities of components with a clear market value are found in many waste streams. For instance, fruit and vegetable phytochemicals may be used to create drugs, cosmetics, or even nutraceuticals. Exploiting such components with the goal of enhancing co-product streams has drawn a lot of attention and research. However, while it is frequently feasible and even very simple to create a procedure to extract a high-value component in a laboratory setting, doing so commercially is frequently unprofitable.

This may be owing to the decreased value of the bulk residues left over after extraction, which may result in higher disposal costs. As a result, it's critical to create strategies that try to utilize food co-products completely, making sure that all components generated may be of marketable quality. This calls for a research and development strategy that connects all potential waste stream components to all possible markets. Utilizing completely traceable, food-grade waste streams is part of this strategy. To maintain their food-grade status, they should initially be stabilized against microbiological degradation and autolysis. They would then be deconstructed using contemporary enzyme and processing technology in physical and biological methods. The goal is to offer a variety of parts, from expensive to inexpensive, each of which would help achieve whole-waste exploitation. It would be necessary to evaluate the procedure for acceptability with respect to novel food legislation, customer preferences, and marketing requirements. To maximize knowledge transfer and exploitation, such a strategy necessitates close communication with all stakeholders. Because of factors like complexity, uncontrolled spoiling, or lack of traceability, some co-products may not be acceptable for use. In such circumstances, the use of non-food products as energy sources through the production of biogas and other microbial-based disposal methods, such as composting, may be appropriate. There may be chances to turn this biomass into biofuels due to new technologies. The perceived return on investment will, of course, serve as the final arbitrator of this strategy's cost-effectiveness, and it will be necessary to consider local and sector-relevant aspects including seasonality, market potential, and transportation and storage expenses. The world is transforming quickly. The drive to cut carbon emissions and the significant rises in fuel prices will undoubtedly have a substantial effect on the food chain. The cost of food processing and transportation is anticipated to rise during the next ten years, along with the need to use more land for non-food crop cultivation. The expensive energy requirements of contemporary intensive agriculture, including the use of fertilizers and mechanical processes, will exacerbate this. These factors, along with the waste hierarchy, will increase the pressure to make sure the majority of food processing and co-products from the food chain are fully utilized. The need for user-friendly LCA is anticipated to grow as more people seek to assess this complicated field from an environmental and economic standpoint (Kluts et al, 2012).

The European Community produces more than 10 million tonnes of waste from the food processing industry each year. Many tens of millions of euros are spent each year on landfill fees and other waste disposal methods to handle the waste generated by the food business. Due to the way food waste is currently processed, it is also recognized that this waste contains substantial amounts of valuable components that go untapped. These wastes contain a variety of potentially beneficial elements, including micronutrients, texturized residues, flavors, and colorants. These could be used as high-value ingredients for medicinal, cosmetic, and nutraceutical goods as well as medium-value ones for food and feed. In the past, the majority of the waste produced by the food processing sector was either used as a landfill or fed to livestock. The food industry has been pushed to reevaluate recycling techniques as a result of recent changes in the legal framework that were implemented in reaction to numerous food safety incidents, such as the bovine spongiform encephalopathy disaster. But right now, the idea of sustainability is receiving more and more attention. As a result, the importance of this issue to the food industry and other food producers is rising. The industry is also aware of the need to react to the growing regulatory, end-user, non-governmental, and consumer interest in the implementation of sustainable manufacturing systems. Consumers are naturally concerned with a variety of issues in addition to sustainability, including food safety, food quality, and emotional reactions to novel foods that may result from food reprocessing. In this sense, customer perceptions of reprocessed food items have evolved into a major force for enhancing waste management in the food processing industry.

4.5 SUSTAINABLE FOOD PRODUCTION

These concerns range from how the public understands sustainability to whether their perceptions align with those of experts in the field of sustainable food production. It's crucial to comprehend how these attitudes and beliefs connect to consuming particular foods and food products. The evaluation of the value of cutting-edge technology and novel foods that may result from recycling food waste must also take into account individual and cross-cultural differences in the conceptualization of and attitudes toward sustainable production processes. The most important consumer perceptions surrounding sustainable and, conversely, unsustainable products and product characteristics, as well as how people's values influence attitudes and opinions, should be identified through research. However, there hasn't been much systematic empirical research done on these problems and challenges up to this point. By reviewing the pertinent literature on connected topics, it may be feasible to generate some predictions about potential barriers to consumer acceptability as well as reasons for customer acceptance, regarding the commercialization of a product. Precise attention should be paid to these concerns when considering the development of new food products from recycled food waste for consumer purchase. Customer acceptance of the many technical advances used in the production process, as well as consumer views of the producers' motivations for introducing these new technologies and perceptions of risk, should all be taken into consideration. Even though the recycled product is not intended for human consumption, previous risk events like the Bovine spongiform encephalopathy (BSE) crisis may have negatively predisposed consumers toward recycling food wastes within the food chain (Frewer & Gremmen, 2007). Consumer anxiety during the BSE crisis was fueled by the recycling of animal waste into the human food supply as animal feed. As a result, consumers may have reservations regarding the utilization of recycled waste in the development of innovative foods.

4.6 STRATEGIES FOR APPROACH

Effective commercialization strategies will put an emphasis on regaining consumer confidence in food production in general, as well as on effectively communicating with and involving consumers in the larger debate about sustainable food production and how this might be achieved and implemented. The elements influencing customer sentiments toward innovative food products will be taken into consideration. The main factor driving the reduction of food chain waste is better sustainability, which is connected to consumer environmental consciousness. Waste minimization and co-product utilization are essential to sustainable development (Sheldon, Bode, & Akakios, 2022). Sustainability literally translates as the capacity to sustain. This suggests keeping something constant or keeping it constant for a while, or offering support by offering assistance and encouragement, or by mobilizing resources. The Brundtland report gave the most popular and well-known definition of sustainable development in recent years, defining it as development that satisfies the requirements of the present without compromising the ability of future generations to satisfy their own needs. Despite the fact that this definition is frequently used, there are several variations, modifications, and reformulations of what sustainability actually is. In fact, sustainability has grown to be associated with a wide range of human endeavors, ideas, and uncertainties, from biodiversity preservation to corporate social responsibility. Furthermore, sustainability has elevated in importance among policymakers. The idea is currently at the center of regulatory operations in many nations throughout the world, and as a result, it has evolved into a cultural standard.

People typically act in accordance with what they perceive to be correct, which may be in accordance with their own personal ideas and/or societal conventions. Many stakeholders, including consumers, believe that the food production system is no longer sustainable. What some refer to as industrial food production or chemical farming systems are the results of the development of more intensive and specialized forms of agriculture. Agrochemical use and dependence, extensive monoculture, mechanization, and energy consumption go hand in hand with it. Such agricultural systems' goal is to generate food and fiber at the lowest possible cost. Although conventional farming practices have historically been beneficial in enhancing food security and safety, many see them as problematic in terms of sustainability. Both the environmental and the social and communal facets of food production are the focus of the issues. For instance, present high-intensity methods in the agri-food sector not only negatively affect water supply and quality, but also create erosion, infertile soils, pesticide contamination, excessive use of synthetic chemicals, and varying degrees of groundwater, soil, and atmosphere pollution. It is arguable that current food production practices may have a harmful impact on the ecology. The food industry is currently making greater efforts to establish more sustainable food production processes. Resources connected to energy and water use conservation are at the top of the sustainability agenda. Many of the objectives of the food industry specifically reflect the need for increased sustainability in the sense of renewable resource use and a significantly reduced environmental burden of food production, despite the fact that sustainability in its broadest sense focuses on the conservation of natural resources.

4.7 FOOD INDUSTRY

The food industry as a whole is worried that consumers won't accept organic by-products as food because they would associate by-products with waste. This would suggest that consumers reject the upgrading of organic by-products given public concerns about, for instance, BSE and other challenges associated with recycling waste materials in the food chain. Consumer rejection of food items would contradict the efforts of specialists in sustainable food production, in addition to the benefits of sustainability not being achieved. The byproduct equals waste scenario is not merely a theoretical possibility. Consumers should play a significant part in deciding what, when, and how unique items are offered. As a result, producers and consumers should communicate more. Negative consumer reactions to the technology push by some producers, most notably the producers of genetically modified foods, have been a significant factor in consumers' opposition to novel agri-food technologies. The reasons why European consumers rejected genetically modified crops have been extensively studied. Public perceptions of agri-food technology, however, are not solely based on an objective evaluation of benefits and hazards. Other elements that affect society and consumer acceptance include moral and ethical issues, as well as other values like respect for the integrity of the environment and confidence in the regulatory system. Creating communication about substantial equivalence failed to allay customer uncertainties and was therefore irrelevant. Additionally, research revealed that European consumers placed a high value on being in control of their consumption of GM foods, calling for their labeling and the installation of efficient tracking mechanisms (Lusk, Roosen, & Fox, 2003).

Figure 4.2. GM foods.

(Source: https://c1.peakpx.com/wallpaper/1/349/264/apples-genetically-modified-foods-corn-wallpaper-preview.jpg, Creative Commons License).

Therefore, the public's unfavorable response to GM foods (Figure 4.2) has less to do with risk and more to do with consumer choice and the lack of information about the real reasons for their concerns. The public was not reassured by opaque risk analysis processes and decision-making procedures. Consumers' confidence in the food industry's motivations for introducing these crops was not aided by the lack of first-generation products with observable and desirable consumer benefits. All things considered, it appears doubtful that consumers will accept the addition of additional GM foods and substances to the human food chain.It is crucial that customers participate in innovation processes in the food processing sector in the future. It's also critical to recognize that customers may act as a storehouse of ideas that might be used during the innovation process far more effectively than is currently the case. In all stages of the innovation process, the creative potential of customers can be utilized in this way to influence technical growth. Of course, the mindsets of those involved in the production process could be a barrier to consumers participating in the design of unique products. Particularly if we are talking about advances like genetics that require at least 5–10 years until the first items get to market, producers may think that consumers don't know what they want. We can only learn about what is being developed through the research and development departments. It is preferable to wait until the first prototypes are out and then consult the consumer because the research is still at an early stage for the public/consumer to be informed. However, given that innovation processes are viewed as dynamic, complicated, and interactive, this reaction might only have a limited amount of relevance. An interactive network model of invention has taken the place of the linear model. In order to shape technology development processes such that social elements are symmetrically taken into account in the process itself, a constructive technology assessment offers to extend the design by taking together involved parties early on and throughout the design process.

Strategic niche management (SNM) is another technique that could aid the innovation process. SNM brings together various stakeholders to test out innovative technologies and discover whether the new technology is desirable through the creation of specific protected zones. SNM accelerates the rate of application and subsequent development of new technologies in this way. To boost customer involvement and influence in the innovation process itself, it is vital to start CTA activities early in the creation of new food items incorporating by-products in order to address this question. Improved food may be used if food byproducts are managed well. Additionally, the consumer will be able to buy food that has functional, health-improving components made from byproducts of industrial processes. These new food items are intended to help consumers get healthier and live better by providing superior nutrition. This can be seen as a component of a larger movement in consumer trends in food consumption. In other words, consumers are starting to pay more attention to the nutritive qualities of food, and they are more conscious of the possibility that consuming certain food products could improve their health. Consumers may, however, view by-products as harmful in the case of novel food manufactured from by-products of the food processing sector, albeit this issue deserves further empirical research. Consumer interest in the management of by-products cannot be assumed a priori, and it is possible that new food products will only be fruitful if customers are aware of their novelty (Meynard, Dedieu, & Bos, 2012). The food business must educate customers about the goals behind product development and the methods used to measure consumer acceptability early in the process. In terms of product development, the manufacturer shouldn't be too "ahead" of the consumers; otherwise, it may be challenging to effectively communicate with customers about new products.

4.8 CUSTOMER TRUST

Consumer trust in food safety is based on a number of well-established variables, which may also apply to other sources of food-related information (Moruzzo et al., 2020). The interactions cannot be efficient and productive without mutual trust. The use of scientific evidence of safety or health benefits is one method the food sector might increase consumer trust. Consumers are unaware of possible advances in this field because there are currently few unique foods on the market that have been created utilizing by-products. This ignorance can widen the gap between customers and producers, which might make it more difficult for them to cooperate effectively. Because some consumers are cautious of novel foods in general or new food products based on novel, unproven technical advancements in particular, it can produce a lack of confidence. The failure of managers, assessors, and other key actors in the process of risk analysis to take into account the actual concerns of the public when assessing, managing, and communicating about food safety issues has led to a great deal of public dissatisfaction with the way food safety issues are managed and regulated. As a result, there is now a greater level of public mistrust in the goals of regulators, science, and business when making choices or taking actions about the priorities for risk assessment, resource allocation, and risk mitigation measures. The public's confidence in food safety continues to erode despite the establishment of new food safety agencies like the European Food Safety Authority(M.-F. Chen, 2008).

A communication's ability to reassure the audience is likely to be restricted if it doesn't directly address their fears. Adopting the fork-to-farm strategy is one way to win consumers' trust. The consumer will be able to buy food that has been produced in an eco-friendly and healthy way due to waste minimization, complete traceability, and the inclusion of Hazard Analysis for Critical Control Points in the manufacturing of co-product ingredients. The hope is that doing so will boost consumer confidence in the food supply, which will be bolstered by the food chain's increased efficiency, which will lead to a cheaper food supply. In addition to an improved quality of life, direct consumer benefits will also include a decrease in the objectionable odors and microbial risks typically connected with the disposal of organic waste. However, a system of information and labeling must be put in place in order to communicate the information to those who need it in order for the customer to generate an enthusiastic response to the new food products.

4.9 RISK

Risks are viewed as greater when people feel powerless to limit their exposure to potential dangers. By providing reassuring results from a reliable source, this effect can be mitigated. Trust in risk assessment and risk management is therefore likely to be a crucial factor in determining public confidence in food safety in situations where there is a lack of control. One example is the issue of GM foods, where customer concern was more about the lack of personal control over intake than it was about risk in general. The public's confidence in the organizations responsible for consumer and public protection was thought to be restored by more comprehensive public engagement and participation in risk management and other science and technology concerns. In fact, this seemed to reflect institutional acknowledgment that attitudes regarding various risks among customers are not solely based on an intellectual evaluation of risk and benefit. Other elements, such as ethical and moral issues, have been acknowledged as having the potential to influence whether or not a specific risk is acceptable or if society will support the steps taken to reduce a specific risk. Although consumer opinions should be seen to influence outcomes and technical developments, failing to do so could have the opposite impact and undermine consumer confidence in risk assessments and technology development and commercialization rather than the other way around.

Consumer perceptions of risk are different from those of other parties involved in food production and risk assessments (Yeung & Morris, 2006). It is commonly known that how people perceive risk affects how they respond to various risks. Some factors can be used to predict how people will react to various hazard domains. Domain-specific factors are also present. People's reactions to a given hazard can be affected by psychological variables. Traditional expert technical risk assessments have minimal impact on the actions and reactions of individuals. People's perceptions of risk, in contrast, are a much more important factor in determining how they react to certain risks. For instance, even when the likelihood of injury is the same or potentially even lower, people perceive involuntary risks as being more dangerous than those they choose to take. Similar to technologically generated risks, naturally occurring risks pose less of a threat. Compared to other comparable hazards with a technological origin, natural risks are seen as less terrifying. People are more afraid of threats that could be catastrophic than ones that could harm a similar number of people at different periods or locations. Other issues are particularly specific to certain risk domains, which is very true in the case of food. It has been demonstrated that public risk perceptions have a significant role in influencing how the public reacts to activities in the agri-food sector. These include biosciences, food safety, and potential unintended negative effects of technology on human health.

All of these instances support the claim that the public's perception of risk has been influenced by the inability to give information pertinent to actual consumer concerns and instead focusing on technical risk estimations generated from expert knowledge. It should be kept in mind that choosing food is as much a cultural, social, and emotional process as it is a rational one, and buying things that are seen as recycled waste may be troublesome due to perceived quality loss. The implementation of useful labeling tactics emphasizing both sustainability and food production is likely to be demanded by consumers as well as the enforcement of efficient traceability systems. The acceptability of the resulting products, particularly in the food industry, may be compromised if consumers believe that food-waste recycling takes place in an opaque manner. Emotional reactions that could be unfavorable to eating food waste could also be a concern. In terms of attitudes, perceptions, and dietary preferences, customers are not uniform, and what is positive or desired by some may be perceived unfavorably by others. Therefore, it's critical to make sure that customers have the option to decide whether or not to ingest foods made with specific technology or for another purpose. Given that the BSE crisis may have impacted consumer acceptance of previous efforts to recycle food waste, it is helpful to think about it from the perspective of by-product recycling (zu Ermgassen,

Phalan, Green, & Balmford, 2016). Technical risk evaluations served as the basis for BSE-related risk communication, which ignored important public safety concerns. Other consumer doubts centered on the use of technology in food production in general and the possibility that these applications could have unexpected consequences. Food processing is an expensive business. Compared to other industries, research and development expenses are quite high, while profit margins are modest. New food items are being subjected to a high and unprecedented level of regulatory scrutiny, which raises the cost of creating innovative products. The creation and expense of such regulations could significantly harm byproduct control in the food processing industry. This suggests that a small number of businesses will control how byproducts are managed.

The corporate takeover of food processing in general is probably going to cause a lot of social anxiety. Despite all the assurances, it seems like the food business still prioritizes short-term objectives when defining agendas. These are related to traditional, high-yield sectors with a focus on profitability. Ownership and control over the food supply chain are directly tied to consumer concerns about corporate dominance. Byproduct management is viewed as an investment that needs to yield a profit. This will spark questions regarding the availability, oversight, and acceptability of using byproducts in food products. Consumers can view the usage of by-products as a technological advancement in sustainability that will result in additional issues/problems. Some customers believe that these developments will lead to unwanted industrialization, economization, and mechanization of agricultural production. Therefore, in order to develop successful commercialization strategies for new, sustainably produced food products, attention must be paid to restoring consumer confidence in food production and food technology as well as to effectively communicating with and involving consumers in the broader discussion on sustainable food production and how this might be accomplished. Understanding what consumers mean by sustainability and how this may be incorporated into product design should be the main focus. Customers should be included in partnerships as potential sources of inspiration for new developments rather than being treated as passive recipients when new food products are introduced. Numerous studies have shown that involving customers in the creation of a product has a favorable impact. This explains why food product innovators that collaborate closely with consumers have a higher success percentage for their innovations. By using such a method, it may be possible to overcome consumer resistance to the use of food technologies in food production. Such consumer dissatisfaction may provide issues for the recycling of food waste, especially if the unique items that result are meant for human or animal use. The food sector may alter its viewpoint on sustainability in society depending on its research goal and results. Future by-product management will only be successful if society acknowledges the direct link between the food industry and sustainable practices. Initial attempts at this might be made on a case-by-case basis, with the outcomes of several cases being utilized to develop generic strategies for the future commercialization of food by-products. Consumers are unlikely to see and realize many of the potential benefits of by-products management unless they can agree that the benefits of by-products management are similar to sustainable, desired, and acceptable food production techniques. For instance, the ISO 14000 certification includes integrated components for quality and environmental policy. Such standards are especially important when companies establish codes of conduct for food safety and/or sustainability. Customers may not be prepared to pay for the development and implementation of the plan due to the resulting increased pricing in the retail sector if they believe that these traits are beneficial to society more broadly rather than to themselves. Consequently, rules of conduct for sustainable production may reflect societal worth rather than customer value, but they shouldn't be linked to higher pricing compared to equivalent goods produced in an unsustainable manner.

Overall, it appears that there is a space and a need for in-depth investigations of the mechanisms of agenda setting in the food processing sector, the level of corporate influence, and the variety of societal perspectives on these topics. Policymakers and scholars appear to understand that it is preferable to incorporate society and earn a license to produce and license to sell than to ignore society and carry on as usual. It is still unclear how this awareness might be used to inform daily management decisions, which calls for additional research and perhaps novel strategies. It is evident that understanding customer concerns and preferences regarding both the food products themselves and the processes by which they have been produced is necessary in order for consumers to embrace innovative foods made utilizing food byproducts. Many corporate managers were inspired by the 1990s environmental movement to consider how their operations might affect the environment. Prior to that, the majority of managers appeared to believe that creating waste was an inevitable part of doing business. Managers are currently placing more emphasis on reducing waste and enhancing waste by-products (Teo & Loosemore, 2001).

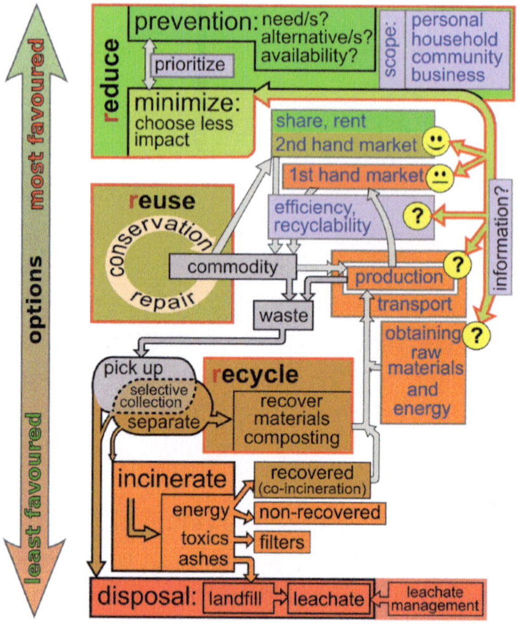

Figure 4.3. Waste minimization.

(Source: https://upload.wikimedia.org/wikipedia/commons/5/5b/Waste_hierarchy_rect-en.svg, Creative Commons License).

4.10 WASTE MINIMIZATION

For businesses in the food processing supply chain, reducing the amount of pre- and post-production organic and inorganic waste is essential. Minimizing inputs, lowering the possibility of waste, and considering re-engineering or producing co-products to add value are essential business goals for minimizing solid waste across the supply chain. Waste minimization is a practical and upfront approach to waste management that can significantly change how a company views and manages waste. Effective waste minimization strategies reduce the amount of waste that is produced, stored, handled, or eliminated by a company. These strategies might range from good housekeeping habits to product redesign. Waste minimization (Figure 4.3) is not only a better option for the environment, but it is also a wise strategy for businesses to use while competing. It enables businesses to boost their profits while lowering rising disposal costs, complying with laws, being seen as a green company, and adhering to government rules. Approximately 222 million tonnes of food waste and byproducts are produced annually in the European Union (Baiano, 2014). An enormous amount of biodegradable waste generated by the food industry disrupts sewage treatment plants and landfills. Biodegradable waste generates methane and causes the release of harmful substances that can contaminate groundwater. The considerable amounts of effluents and residues discarded by food processing companies can overload wastewater treatment systems with biological and chemical oxygen demand. Excessive packaging, waste oils, food waste, out-of-date stock, empty raw material containers, damaged pallets, used filters, and sludge are examples of solid waste produced by food processing facilities. All of these solid wastes amount to a waste of money and resources. Avoiding the creation of waste in the first place is the most efficient strategy to reduce the losses related to waste. The production of less solid waste will result in cheaper operational expenses, lower waste disposal costs, improved workplace health and safety, decreased long-term liability, support environmental quality maintenance, and favorable public perception. Solid waste has become overabundant due to excessive packaging.

Food processing industries are rethinking their usage of packaging due to rising solid waste disposal prices, dwindling landfill space, and customer pressure. Packaging modifications are also being impacted by the legislation. The European Union has deemed it unsustainable to dispose of garbage in landfills. The final choice in the hierarchy for disposing of solid waste is to landfill. Landfilling is regarded as unsustainable for a variety of reasons. Landfills occupy valuable land and pollute the soil, water, and air. In addition to releasing CO_2 and methane into the atmosphere, they also release harmful compounds into the soil and groundwater. From the farmer to the retailer, a network of businesses makes up the food supply chain. A company must cooperate with its suppliers and distributors in order to meet customer needs more effectively and efficiently. Although it has taken the food sector a while to achieve the same success in chain management optimization as the automotive industry, it seems that it is increasingly gaining importance across the entire global food industry. Competitive trading relationships were replaced by ones based on cooperation and coordination, and businesses started sharing operational and strategic information. Chain management solutions are helping food producers save costs significantly by shortening times for production, lowering inventory levels, and speeding up the development of new products. Economic pressures are causing the global agri-food chain to change and are encouraging more vertical and horizontal cooperation to create joint marketing partnerships. It improves the performance of the entire supply chain and boosts its efficiency and responsiveness by forging deeper relationships between suppliers at all stages of the chain. A partnership built on trust between all of the participating parties is essential to supply-chain waste reduction programs. This is different from how business is currently conducted, which is competitive, and from how the best price for goods and services given throughout the supply chain is negotiated. It is crucial to create a cooperative culture and secure a strong union among supply chain participants because of this. Environmental supply-chain management places an emphasis on practices that minimize waste and maximize resource efficiency. The ideas of cooperation and integration should be applied jointly to reduce waste. Waste control efforts within the supply chain for the food processing industry must be coordinated by all participating businesses. It is important to consider each product's environmental effects over its whole life cycle. Avoiding food waste entirely is the greatest approach to reduce it.

This can be facilitated by streamlining processes and using resources more efficiently. Since reduced product/raw material purchases result from increased efficiency, it is difficult to allocate cost benefits across supply-chain participants that practice waste minimization. In some ways, the affected businesses encourage the inefficiency of supply chains because it boosts their revenues. Even with the obstacles standing in the way of supply-chain collaboration on waste reduction challenges, progress is likely to be made as the business community accepts joint accountability for the environmental effects of its products. It is the first and most crucial step in reducing waste and conserving resources. Due to its simplicity and low cost, good housekeeping is typically at the top of the list of waste minimization techniques. By following standard routines, effective housekeeping aims to reduce material losses and stop the creation of unneeded waste. Good housekeeping procedures are similar to those in a home in commercial settings. They are actions that may be taken to preserve cleanliness and avoid spills or mishaps. Enhancing maintenance processes, keeping an eye on leaking valves and fittings, ensuring correct material handling and transfer, separating waste streams, and strengthening inventory controls are all examples of good housekeeping techniques. All operations in food processing facilities, including purchasing, materials accounting, inventory management, receiving, delivery, storage, correct labeling, optimizing operations and processes, preventative maintenance, and recovery/reuse/recycling, can benefit from good housekeeping standards.

4.11 PACKAGING DECISIONS AND APPROACHES

There are numerous strategies to cut down on packaging waste. Cardboard boxes, pallets, plastic-lined cartons, plastic bags, plastic containers, plastic film, metal cans, paper, and aluminum foil are just a few examples of the packaging waste generated during shipping and receiving operations. Make the decision to buy items in less packaging and in bulk. For items that are utilized frequently, purchase larger sizes. Find ways to internally repurpose some of the packaging. Make arrangements with suppliers and clients to utilize returnable packaging, less packaging, easily recyclable packaging, and packaging with recycled materials. Send back empty containers to the source for recycling.

Overproduction, product deterioration, broken items, spillage, or operator errors are the causes of this waste. The most effective way to reduce this waste stream will be to improve process control because inaccurate weights and contamination are the main causes of waste. Waste can be reduced by thoroughly draining mixers, troughs, and tanks prior to cleaning them, as well as by removing sediments from machinery or the floor before soaking such surfaces. Waste from bakeries can be composted or fed to animals. This waste consists of unfinished goods, rotten materials, sauces, grease, spices, additives, oils, and drippings.

Recycling leftover grease, cooking oil, and pork fat will help reduce waste by preventing dehydration and spoiling when storing raw veggies in reusable containers. Fluid milk and processed milk products are the two segments that make up the dairy industry. By designating lines for specific products, avoiding product spillage when removing hoses and pipes, offering a foolproof whey collection system to prevent leaks from valves and fittings, offering whey storage tanks with a capacity twice as great as the maximum daily volume to prevent tank overflow, and preventing sludge from entering the wastewater stream, it is possible to reduce solid waste. Improving the quality of the feed that is delivered to the plant, using vacuum to remove skin, fat, and flesh, installing trays around the machines to catch solids, vacuuming for processing oily feed are all ways to reduce waste. Waste can be composted, utilized to make pet food, protein hydrolysate, fish meal and oil, or recycled into these products. Odor and noise may also be problems in some locations. Improved staff housekeeping habits, process optimization, the adoption of new technology, and/or new product design are all strategies for reducing waste and increasing operational efficacy. In fact, the processor can benefit from lower operating costs, lower waste treatment/disposal expenses, and lower liability through successful application of various approaches meant to reduce inefficient use of resources and eliminate unnecessary waste formation.

4.12 WATER MANAGEMENT

Large amounts of water are used in the slaughterhouse environment for the watering and washing of cattle, washing of transport infrastructure, washing of carcasses and by-/co-products, cleaning and sterilizing equipment and processing spaces, and washing of carcasses (de Nardi et al , 2011). Technical fixes or equipment upgrades might be used as water consumption reduction strategies.

The majority of water used in abattoirs eventually turns into effluent. The high quantities of organic matter in abattoir effluent are caused by the presence of dung, blood, fat, and other animal remnants.

The effluent may have significant concentrations of salt, phosphates, and nitrates. Blood and fat-containing materials are the two most significant sources of the organic load, and blood is the main source of nitrogen in the effluent stream. High levels of sodium and phosphorus, which typically come from manure and stomach contents, can also be found in effluent. The effluent often represents the single most significant source of pollution load from the abattoir at those operations where rendering of animal remains takes place. Water is heated at a slaughterhouse (Figure 4.4) to be used for cleaning, to remove hair, blood coagulation, and blood drying.

The remaining 20% of the total energy used by abattoirs is typically provided by electricity, which is used to run equipment in the slaughter and boning areas, for processing by-products and co-products, for refrigeration, and for compressed air. Through simple housekeeping techniques, significant energy consumption savings can be achieved almost immediately with minimal capital outlay. There are a variety of worker and process/product enhancements that may be made to reduce waste and increase processing efficiency in a contemporary abattoir and meat processing plant. These include routine maintenance and work practice improvements, resource-saving optimization of current processes and equipment, adoption of new technologies to cut down on resource use and minimize waste through improved operating efficiencies, and modifications to product design to lessen waste generation and treatment, use less energy, and increase production efficiency.

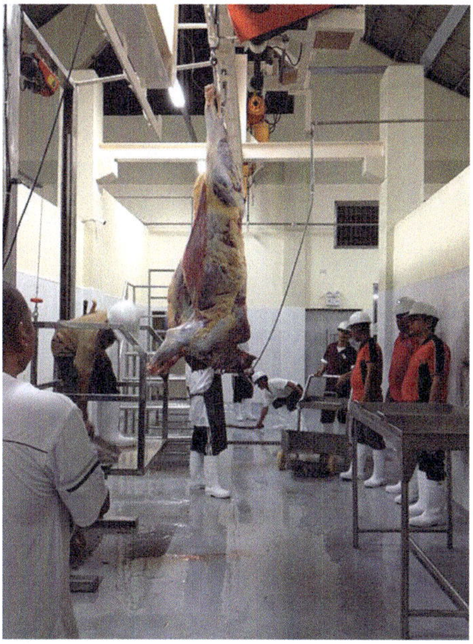

Figure 4.4. Slaughterhouse.

(Source: https://upload.wikimedia.org/wikipedia/commons/thumb/1/11/Cotabato_City_Halal_Slaughterhouse.jpg/450px-Cotabato_City_Halal_Slaughterhouse.jpg?20180504204855, Creative Commons License).

Prior to their separation and stabilization, these raw materials are stored for extended periods at room temperature, which increases the generation of offensive odors and lowers the value of the resulting protein meals and fat fraction. Rapid processing enhances final product quality while minimizing offensive odors. Biological scrubbers work by microbially breaking down air pollutants that have been absorbed in the scrubber media. An easy and efficient way to significantly cut objectionable odor emissions is to burn gases in an existing boiler. A heat exchanger is used to dry out the damp air. The residual odor-filled air is burned while the condensate is sent for wastewater treatment. Water reusing might significantly reduce water use and outflow by the food sectors, especially the meat processing sector. However, because of a number of limitations and obstacles, water reuse has not been widely used. In the past, meat manufacturers have consolidated all of their operations' wastewater and sent it to a single treatment facility. More recently, meat processors have realized the advantages of separating different process water streams in order to maximize water treatment or allow for direct re-use. Due to the high temperatures, the majority of sterilizer water is only marginally contaminated and pathogenic microorganisms are absent. Reusing cold water from beef processing, and carcass washing water from poultry processing are all examples of improved water treatment (Kist, Moutaqi, & Machado, 2009). In particular, the European Commission has gathered a sizable amount of expertise addressing the waste challenges affecting the meat processing industry. The removal of foreign suspended and/or dissolved solids, organic substances, and inorganic salts are all part of the treatment process for wastewater. This elimination is frequently accomplished through a hierarchy of treatments. These treatments are divided into primary, secondary, and tertiary categories. Mechanical screening, fat separation, dissolved air flotation, or coagulation/flocculation may be used as first treatments. Secondary treatment often entails a biological process that turns colloidal and soluble materials into biosolids. You can utilize aerobic digestion and anaerobic digestion separately or together.

4.13 PROTEIN PROCESSING

It has been suggested that the blood plasma of cattle makes a useful nutritious addition to meat products. Blood proteins are used as a protein source in the pet food business as they provide all the essential amino acids for a carnivorous pet diet. The demand for blood proteins from the pet food sector is predicted to rise in the upcoming years. The pet food market is quickly expanding in Europe. Due to its ability to gel, dehydrated blood plasma is also utilized as a protein source, particularly in meat derivative products. Blood is often concentrated by evaporation or ultrafiltration at low pressure. Their gelation properties can be modified based on the resulting pH content. Red blood cell concentrate, blood plasma protein concentrate, globin isolate, and a carboxymethylcellulose-heme-iron complex have all been made from fractionated bovine blood. It has been established that the plasma protein concentrate's amino acid composition is well balanced, resulting in net protein utilization and a net protein ratio that is 95% as high as casein. The solubility of globin at neutral pH has been observed to be minimal, and it has advantages over other proteins used in the food sector. However, by hydrolyzing globin with citric acid, its useful qualities can be enhanced.

Refined animal protein isolates are made from non-blood sources, like the collagen-rich rind of pigs (Roupas, De Silva, Smithers, & Ferguson, 2007). The isolation of intramuscular collagen while maintaining intact fibers, has been achieved by ultrasonic fragmentation of myofibrils and subsequent solubilization in the buffer. Many meat protein isolates have a protein content of above 90% and are utilized in hotdogs in place of carrageenan because of their ability to emulsify. Smaller amounts of these goods are made from chicken, with the majority originating from pork. Bovine skin naturally contains collagen, which can be transformed through the correct processes into a very thin collagen film with edible qualities. Such a collagen film can be utilized to both enclose the sausage meat and act as a separating or releasing layer between the sausage meat and the net when making uncooked sausage that is net-wrapped. Researchers have examined how the amount of collagen affects the biological value of beef proteins using computer-aided simulation. It has been demonstrated that the protein utilization for tissue synthesis was significantly improved by increasing the collagen content of powdered beef from 2.5 percent to 15-20 percent of the total amount of proteins. Meat products with the aforementioned collagen content have higher nutritional and biological value and are better suited to human metabolism. It has been claimed that collagen, soy protein, and carrageenan work together to improve the texture and water-holding ability of rolls. Collagen proteins have been employed to help enhance the technical and sensory qualities of meat products and prepared foods (Hjelm et al, 2019).

Collagen content has also been utilized as a gauge for the quality of the raw materials used in the production of meat products. According to reports, large levels of collagen and, consequently, connective tissue, which are discovered in meat samples, are good indicators of the poor quality of the raw materials used to prepare meat products. Collagen is also a good indicator of protein quality. Physical functionality has been found to be affected by freeze-drying collagenous material from chicken foot skins and tendons because it increases solubility, and water-holding capacity at 60 °C. The potential use of these products as functional components in meat products would be indicated by such modifications to this co-product of the processing of chicken. A protein product called gelatin is commercially produced from pork skin, beef bones, calf skin, or fish skin. It is made by thermally denaturing collagen that has been extracted from animal bones and skin while using a very weak acid. Numerous amino acids, including glycine, proline, and 4-hydroxyproline residues, are present in gelatin. Gelatin is a highly digested food protein that can be added to foods to boost their protein level. Gelatin is a heterogeneous collection of water-soluble proteins made from collagen material. In terms of functioning, gelatin is arguably the most adaptable of the animal proteins. Gelatin is used in many different products as an ingredient, including baked goods, dairy products, meat products, pet meals, and pharmaceuticals. Gelatin is added to confectionery items to improve the foaming, gelling, and final solidification of the candy so that it dissolves gradually when consumed. It provides qualities like texture and prevents the sugar from crystallizing in some products, like marshmallows.

Chapter

5 SUSTAINABLE FOOD WASTE MANAGEMENT

Content

5.1 INTRODUCTION

Food production and consumption habits are significantly impacted by the development activities and changes in the current lifestyle. Because of changes in demographics, changing lifestyles, increased per capita income, and shifting moral and social standards, society's attitudes about food have changed over time. The diversity and accessibility of food products have changed as a result of technological advancements and competitiveness in the global food market. However, the issue of affordable and accessible food is just as relevant today as it was decades ago.

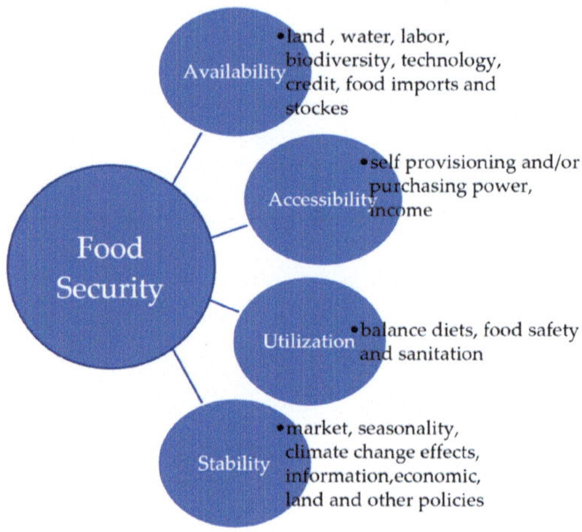

Figure 5.1. Food security.

(Source: Charoenratana S, Anukul C, Rosset PM. Food Sovereignty and Food Security: Livelihood Strategies Pursued by Farmers during the Maize Monoculture Boom in Northern Thailand. Sustainability. 2021; 13(17):9821. https://doi.org/10.3390/su13179821, Creative Commons License).

Food security (Figure 5.1) is in danger due to careless behavior both now and in the future. The predicted 60 percent rise in global food consumption by 2050, the consequences of climate change, the scarcity of natural resources, the loss of yield and land area due to environmental degradation, and competing demands, particularly for the development of biofuels make this clear (van Dijk, Morley, Rau, & Saghai, 2021). 60 percent of the major ecosystems are currently damaged or being utilized in an unsustainable manner (Islam, Sunny, Hossain, & Friess, 2018). Global population expansion will also increase the number of wealthy individuals, which will increase the need for resources to create goods for their diversified, high-quality diets. Food that is produced and cultivated but not consumed has a major financial and environmental impact. It has a direct and detrimental effect on the income of both farmers and customers and results in the waste of resources utilized in production, including fertilizers, energy, water, and land.

The specific reasons why food is rejected rely largely on the circumstances and local scenario that a country is experiencing. For example, in low-income nations, these issues are associated with financial, managerial, and technological constraints in harvesting methods, storage and cooling facilities in challenging weather conditions, infrastructure, packaging, and marketing systems. In contrast, in middle- and high-income countries, the causes are related to consumer behavior, quality standards, legislation, a lack of coordination between various supply chain performers that results in oversupply and overproduction, technical issues, and difficulties predicting consumer demand. The retail business model sees food disposal as a necessary aspect of the operation. According to the World Economic Forum's 2009 report, the harvesting, storing, shipping, and processing phases account for 35 to 50 percent of food loss or waste in developing nations (Wunderlich & Martinez, 2018). While a consumer in Europe and North America typically discards 95–115 kg annually, that number is merely 6–11 kg in South and Southeast Asia and sub-Saharan Africa. In the US, 40% of the food produced is wasted. Compared to American consumers in the 1970s, the average American today wastes up to 50% more food. The majority of food-related waste in the EU is produced by homes, accounting for roughly 42 percent of the total or about 38 million tonnes. According to research done at the national level in the United Kingdom, 4.1 million tonnes, or 61%, of the food produced is wasted because it was not properly managed. Fresh, raw, or barely processed food makes up 46% of the food that is wasted, whereas heated or prepared food makes up 27% and food that is ready to eat at the time of purchase makes up 20%.

Nevertheless, despite the fact that a number of studies have been conducted that support addressing the issue, actual precise data on food loss and waste generation, avoidance and management and its ultimate fate are rare, fragmented, disaggregated, and challenging to verify, both on the global and national levels. As a result, the information on the amount of food waste produced in each nation provides a basis for proving or disproving the idea that there is a strong negative correlation between the amount of household food waste and the share of food expenditures, i.e., that consumers who spend less on food produced more food waste. 11.2 billion tonnes of solid waste or SW are collected annually around the world (Ayodele, Ogunjuyigbe, & Alao, 2017). Along with population growth, urbanization rates, GDP/GNI per capita growth, increases in production and consumption, and changes in consumption patterns, the amount of accumulated waste will continue to rise in the future years. In addition, the most recent World Bank analysis projects that by 2025, the cost of managing solid waste globally will rise from USD 205.4 billion to around USD 375.5 billion (Yokoo, Kawai, & Higuchi, 2018). The enormous economic potential of this waste, however, is a plus. The projected value of the global waste market today, from collection to recycling, is USD 410 billion annually, excluding the sizable informal sector in developing nations. Such a large volume of waste and its complexity not only threaten human health, pollute the environment, and emit greenhouse gases, but they also waste a tremendous amount of material and energy resources. Europe, which is heavily dependent on imported raw resources, aims to reduce the amount of waste it produces by enhancing resource efficiency through recycling, waste avoidance, and the use of unavoidable waste as a resource whenever possible.

5.2 LANDFILL MANAGEMENT

The EU Landfill Directive is another significant directive that outlines the fundamental conditions for waste disposal. The Directive establishes maximum landfill site capacities and aims to reduce the amount of biodegradable municipal waste dumped in landfills. It prohibits the placement of specific waste streams in landfills. Each member state must develop a national strategy for activities including recycling, composting, recovery, and biogas production that aim to reduce the amount of BMW. It includes specifications for starting and managing a landfill throughout its operating and after-care phases. The examination of accomplishments in this area, however, reveals that in 2010, despite notable gains in material recycling, the majority of European countries continued to send more than half of their municipal waste to landfills. Additionally, it requires the member states to set up systems for energy recovery, return, and collection, as well as to promote the use of products made from recycled packaging waste. The best way to stop the production of packaging waste is to reduce the volume of packaging used overall. As per the UK Department of Energy and Climate Change, the present framework is a five-step hierarchy of waste management and waste treatment solutions arranged in order of what is best for the environment. It is a set of guidelines for planning waste management, collecting waste, and treating that waste. A framework like this is useful for comprehending how management strategies might affect materials as they move through the material life cycle. However, the hierarchy undergoes modifications in each unique situation. Diverse specialist organizations and institutions modify the waste hierarchy by enlarging or constricting the content of its stages with waste prevention as their ultimate objective. The US Environmental Protection Agency is in charge of carrying it out in the country. The major regulation that controls how solid and hazardous wastes are disposed of in the nation and is enforced by the EPA is the Resource Conservation and Recovery Act. The Integrated Solid Waste Management approach makes use of a framework called a waste management hierarchy (Geng, Zhu, & Haight, 2007).

The Waste Framework Directive (WFD) establishes energy efficiency limits beyond which the incineration of MSW may not be recognized as recovery in order to discourage the combustion of bio-waste with low calorific value. The WFD mandates that it promotes the use of biodegradable materials and takes into account potential biodegradable waste composting and digestion possibilities through separate collection. Separate collection has several advantages, such as making it simple to divert biodegradable waste from landfills, increasing the calorific value of the remaining MSW, and creating a cleaner bio-waste fraction that makes it possible to produce high-quality compost and streamlining the production of biogas. Some have already put programs in place for separate collection, diverting biowaste from landfills, and preventing bio-waste through the use of economic instruments or targets. However, no national bio-waste-specific prevention targets were found. Consequently, the EU regulation does not restrict Member States for options for treating biowaste. The Commission advises using the Life Cycle Assessment tool to plan the management of bio-waste in order to overcome these challenges and support decision-makers in making the optimal use of biodegradable waste in accordance with the waste hierarchy. Such a strategy can be utilized in conjunction with the waste hierarchy to ensure that the best overall environmental alternative is found. The formation of high-quality compost and biogas helps to improve soil quality, resource efficiency, and energy self-sufficiency.

Despite the aforementioned restrictions and obstacles, the Commission's actions, such as legal limitations and support for a range of programs, projects, and initiatives, are having a good impact. The management of bio-waste could assist the environment and the economy by being in line with the WFD's rules and waste hierarchy. The financial benefits could range from EUR 1.5 to EUR 7 billion as a result of the announcement from the Commission to the Council and the European Parliament on future steps in managing bio-waste in the EU, and as a result, environmental benefits could include an estimated 34 million tonnes CO_2 equivalent saving, of which 80–90% would be attributable to prevention.

5.3 FOOD WASTE

Food waste causes a number of adverse environmental effects and has significant financial implications. Food waste contributes to climate change by increasing greenhouse gas emissions unnecessarily and wasting water and land. As a result, natural ecosystems and the services they offer begin to deteriorate. Economically, food loss and waste totals over USD 310 billion in underdeveloped nations and USD 680 billion in affluent nations (Ananno et al., 2021). It represents wasted expenditures that may lower producer incomes and raise consumer costs. Ethically, it means that possibilities to feed the expanding global population are lost. It is possible to prevent one-fourth of the food that is currently lost or wasted worldwide, which would be sufficient to feed the 870 million hungry people throughout the world. The European Commission has identified food waste as a significant household waste stream that must be avoided; as a result, any biodegradable waste -prevention activity must focus on reducing food waste, and the support of such activities at the EU level would have the greatest impact.

Food was not defined to include inedible parts like skins, stalks, leaves, and seeds, potential foods, and feed. Loss was described as any modification to the food's availability, palatability, freshness, or quality that precludes people from eating it. The main issue is avoidable food waste since it involves food that, with better management or storage, could have been consumed. According to the Water Reform Action Plan (WRAP) and the NSW Office of Environment and Heritage, Australia, the food may have been otherwise spoiled at the time of disposal, rendering it unfit for human consumption. The European Parliament Resolution addresses key aspects of the issue of food waste and lists causes of food waste such as overproduction, improper product targeting, deterioration of the product or its packaging, marketing rules, and insufficient stock management or marketing strategies. Both wealthy and developing nations experience food losses and waste in the supply chain for food. In rich countries, the bulk of food waste is produced at the shop and consuming stages, but in developing countries, over 40%

of food loss/waste emerges during the production, harvest, processing, storage, and transportation stages. When compared to sub-Saharan Africa and South and south-eastern Asia, consumer waste per person in Europe and North America is between 95 and 115 kg per year (de Lange & Nahman, 2015). The resolution of the European Parliament outlines a list of necessary actions at the European level. The document urges, among other things, that the issue of food waste along the entire supply and consumption chain be addressed immediately, as well as that sector-by-sector developments in the efficiency of the food supply chain be made, as well as an analysis of the effects of waste, disposal, and landfilling, as well as the resulting economic, environmental, nutritional, and social impacts, be made.

The European Commission has prepared specific guidelines to address food waste in order to assist decision-makers in creating National Waste Prevention Programs. The plans on the preparation of food waste prevention programs address such important sectors as households, the hospitality industry, the retail supply chain, businesses and institutions and provide a general approach to food waste prevention as well as guidelines for developing a food waste prevention program. According to several studies, consumers are the main source of food chain waste. The findings of the UK study show that 60 percent of all food thrown includes food waste. More than 25% of this waste still has its original packaging on it. Food waste represents a substantial percentage of the municipal and residential waste streams. Factors contributing to household food waste are excessive preparation, preparation waste, plate waste, spoiled leftovers, breakage, spillage, and package failure, either at home or while being transported from the place of purchase. The reasons for food waste in households include a lack of consumer education regarding food safety, inadequate knowledge for interpreting package date information, expiration codes, as well as confusion about an influence of quality defects on edibility. The total amount of food waste generated is significantly influenced by consumer preferences as well. Customers do want as many options as possible, therefore they anticipate finding a variety of goods in stores at any given time. Food waste generation is also influenced by cultural factors including the practice of wasting food to show off one's wealth or the practice of just consuming specific food components. There are other factors than consumer preferences and a lack of information that contribute to food waste. High customer expectations are a contributing factor in some situations, notably in industrialized nations. One of these causes is the aesthetic requirements for products. Food that is totally safe and suitable for ingestion may be rejected merely due to a physical or cosmetic flaw, such as having the incorrect shape or size, being broken, etc. The fixation with cosmetic quality standards has nothing to do with how healthy food is. Supermarkets establish high-quality requirements for weight, size, form, and appearance to match consumer expectations. Due to these false standards, a sizable amount of perfectly edible fruits and vegetables are rejected by retailers. For instance, as a result of such methods, up to 30% of the vegetable production in the UK is never harvested (Barbieri, Pellerin, Seufert, & Nesme, 2019).

Culling is the process of removing products based on quality or appearance standards, such as requirements for size, color, weight, degree of defect, etc. Moreover, trimming is frequently done on food processing systems to guarantee that the final product is the proper size and form. Typically, these trimmings are thrown away. However, producers choose to leave undersized, deformed, or otherwise imperfect fruit in the field during the harvesting stage, as these products would probably be thrown away in the packing shed or processing factory, even before this point. Technology also contributes to the production of food waste. Due to the aforementioned aesthetic standards, errors in food processing that result in final products with the incorrect weight, shape, or appearance, or damaged packaging, without altering the safety, taste, or nutritional value of the food, cause such products to be discarded at the production stage. Any marketing strategy's primary objective is to boost sales. Food product promotion activities are not an exception. As a result, marketing as a technique significantly contributes to encouraging careless consumer behavior. Large amounts of food are wasted at home as a result of promotions like "buy one, get one free" coupons, super-sized servings, bulk discounts, and high-pressure marketing campaigns. By shifting the responsibility for the additional food waste produced to the consumer, these kinds of marketing operations can be seen as a means of preventing food waste in addition to being a tool to increase profit.

Notwithstanding the wide variety of products available today, it is nearly impossible to find a product with the necessary container size. Typically, this occurs because businesses infrequently consider the various dietary and energy needs of their clients, including men, women, and children. The food service and hospitality sectors also experience a similar issue. Restaurants, canteens, and other establishments encourage customers to leave leftovers on their plates since some people are unable to finish a meal by just providing single options for portion size or buffet. Legislation appears to be both one of the barriers to reducing food waste and one of the solutions to solving the problem of food waste. Legislative or regulatory obstacles typically arise when treating collected food waste. There are instances where customers and/or distributors attempted to prevent food waste but were unable to do so due to restrictions, primarily those pertaining to food safety. For instance, a town in France forbade a school from redistributing the bread that was left over from lunch. Due to a provision in the French Environmental Code that holds the store owner liable in court if someone who obtains food waste becomes ill, stores also forbid the consumption of food that is designated for disposal. The Product Liability Directive governs similar cases at the European level. The agreement states that a producer, importer, or supplier may be held legally accountable when a defective product harms a customer. Therefore, businesses believe that a discard alternative is the safest. In the Netherlands, businesses view the harsher cleanliness regulations than the laws upon which they are founded as a legal barrier that leads to needless food waste. The actors occasionally go too far when discussing the subject of food safety, and the hygiene rules contain very broad safety margins. On unrefrigerated products that are offered for sale, there are additional time restrictions, such as "the 2-h guarantee." According to the rule, goods that typically require refrigeration can only be sold for a maximum of two hours before they must be thrown away, whether they are packaged or not. The European laws outline both basic and specialized marketing requirements for certain food product categories (Murdoch & Miele, 1999). The quality, grading, tolerance, packaging, and name are all governed by distinct marketing guidelines.

5.4 MARKETING GUIDELINES

One of the perceived reasons for food waste in the food chain is the European Marketing Standards for fresh fruit and vegetables. The purpose of marketing standards is to ensure EU consumers a certain minimum level of product quality. They are mostly concerned with the product quality and labeling of a retail stage. The rule establishes the broad and detailed marketing requirements for Fresh Fruit and Vegetables (FFV). The modifying version also eliminated rules that specified the precise shape, size, and appearance of fruits and vegetables. Research works continue to outline product-specific standards for each of these 10 goods, including categorization, minimum size and quality, sorting, and maturity requirements. Products that don't conform to marketing norms might not be sold as fresh goods within the chain. The marketing and commercial quality control of FFV, including the 10 varieties covered by the particular EU marketing standards, have also been standardized by UNECE (United Nations Economic Commission for Europe).

Individual product-specific EU marketing requirements must be consistent with applicable UNECE requirements. Once a year, these criteria are modified. Many EU purchasers prefer the UNECE standards because they are stricter than the EU norms (Leal Filho & Kovaleva, 2015a). The removal of a few marketing criteria didn't have a significant impact. In EU supermarkets, it is still challenging to find fruits and vegetables with unusual shapes. The reasons include the use of outdated standards as private quality requirements, the slow adaptation of the current supply chains to new opportunities, and the continued sensitivity of EU customers to a certain FFV shape and appearance.

5.5 FOOD CHAIN OPERATIONS

Some chain operators in the fruit and vegetable trade today have adopted the old original European marketing rules in the form of a proprietary classification system. One of the primary causes of this behavior is a company's reputation, which directly affects its profitability. Companies work very hard to maintain and defend it as a result. Companies in the food business establish private standards that are more stringent than those mandated by law in order to prevent legal action and safeguard their brand at the same time. The technology, which is dependent on the quality standards, is another justification for private standards. Standard processing lines are typically suitable for goods of the same size and shape. Studies conducted in the Netherlands revealed that, for instance, a fruit grower wishes to employ as many "A class" products as possible to minimize losses in an operation since processing different forms or leftovers is ineffective because the equipment is not designed for them. The same holds true for logistical processes. Vegetables with non-standard shapes are typically not sold in conventional supermarkets since it is far more efficient to pack goods with standard sizes and shapes than with non-standard shapes. Other private standards include those relating to delivery temperature, lower Maximum Residue Levels (MRLs), higher legal levels of pesticide residue concentration in food, and expiration dates. Additionally, they are more stringent than what is mandated by law. For instance, businesses frequently demand a specific expiration date for fresh products. If the "best before" date is too soon, people won't buy the product. Another illustration is when grocery stores utilize 50% of the permitted MRLs as their norm because measurement errors might be up to 50%. As a result, fewer crop protection chemicals are utilized, and putrefaction-related waste production increases. A consortium of EU retailers and supermarkets has created additional standards for FFV like Global GAP to ensure a minimum level of safety in response to growing customer concerns about food safety. The biggest retailers and wholesalers in the EU demand the Global GAP certification. Some supermarkets, including Tesco in the UK and large supermarkets in France, have created their own firm-specific requirements.

The biggest cause of food waste in the Netherlands, according to primary producers, is the overproduction of fruits and vegetables. There are several causes of overproduction, including disagreements over the appropriate production levels that lead to a substantial discrepancy between supply and demand. Additionally, it has an economic impact. More products provided to the market results in declining prices; consequently, fruits and vegetables are occasionally destroyed to avoid such scenarios and maintain the price artificially high. In reaction to the uncertainties imposed by retailers, there is purposeful overproduction in the food manufacturing sector. If a retailer makes a last-minute alteration to an order, the supplier must pay the additional or reduced amount of items required or peril being removed from the list of suppliers. Bakeries manufacture more than is necessary merely to give the appearance that their products are available. Unnecessary excess stock is created by one of the actors in the food chain as a result of poor inventory management, as seen in inaccurate ordering or demand forecasting, the "take-back" system, and last-minute order cancellations. These factors are due to supply contracts that require retailers to return stock to their suppliers once it has reached a certain amount of remaining shelf-life. As a result, suppliers have few choices for getting rid of the returned inventory. Due to this, large amounts of food waste are dumped rather than sold without risk to human health. According to published studies on supermarket waste, retailers also discard food as a result of overstocking or insufficient stock rotation of seasonal commodities like Easter eggs (Leal Filho & Kovaleva, 2015b).

According to research findings from various EU nations, producers and importers are unable to predict their sales volume during the planning stage. They overorder or overproduce, which results in food waste. While supermarket companies provide sales estimates, they do not accept any liability for the information. Suppliers place orders or produce according to these projections, but supermarkets may only purchase a small portion of the quantity predicted; the remainder is wasted food. Companies are demotivated to lower their existing levels of food waste since there aren't enough financial incentives. This also applies to residual fluxes. Companies won't explore for other methods of recycling their residual flows if they produce no income from them and/or can dispose of their waste for a low cost. The processing sector also cites a lack of market demand, a lack of funding for enhanced waste flow value utilization, and a reluctance to invest in waste flow separation because the expenditure would not pay for itself as a barrier to reuse waste flows. Grocery retailers are not offered any financial incentives to lower waste levels. One of the factors behind this is that disposing of excess supply is less expensive than other choices in terms of cost. Price is another crucial consideration. Crops may not be harvested in some cases simply because the price is too low and does not pay for the labor and transportation necessary to pick and sell that product. Because of this, it is financially sensible for a farmer to let that food go to waste. The caterers are not necessarily financially disadvantaged by food waste. 40% of catering contracts include payment for waste removal. Therefore, waste is still a customer's and not a caterer's property. It is also important to note that the European Parliament works to reduce food waste in order to lower food prices. On the one hand, it will boost the population's ability to acquire food, but on the other, it may result in a rise in food waste due to the rebound effect. Food goods that are damaged as a result of poor packing performance will be thrown away. When a product's primary or secondary packaging is damaged, even when the food inside is undamaged, the product will frequently be abandoned. Packaging is occasionally rejected and frequently destroyed because its contents are not homogeneous owing to production faults and disruptions. In addition, incorrect packaging presents a technological difficulty for the unpacking of packaged goods because it takes a long time, involves a lot of work, and is quite expensive.

5.6 FOOD BANKS

As a result, each of the multiple factors that contribute to the formation of food waste is influenced by both other factors and actors outside the food chain. For instance, inaccurate sales estimates, overproduction, oversupply, and food disposal are all actions that are prompted by consumer expectations about a product's availability and lead to food waste. This wide range of issues and impacted regions suggests that the problem requires the participation of numerous actors from all links in the food supply chain. The main players in the food supply chain should decide on waste reduction strategies that focus on social, cultural, economic, and legal spheres. Better education, increased public awareness of the problem's status, its effects on the environment, strategies for preventing food waste, etc., could help realize the actions, programs, and initiatives that are aimed at altering attitudes and behavior toward the problem of food waste. The cultural perspective of food must also be changed, i.e., the assumption that low-value food types are abundant and depend on aesthetically pleasing standards for quality must be replaced. A lot of attention must also be paid to the problem of food labeling, as well as to a better interpretation of the meaning of the label content (Madilo et al, 2020). Application of already known technologies and the creation of new ones also aid in the problem resolution, particularly during the stages of harvesting, manufacture, distribution, and marketing. However, it is believed that the main forces behind the implementation of measures aimed at reducing food waste at both the national and international levels are governmental interventions through laws, regulating policies, and economic incentives, as well as non-governmental initiatives.

Training programs cover topics such as teaching consumers waste-free cooking techniques, using leftovers, training staff about the effects of food waste, and teaching workers how to avoid it while at work. Because social institutions occasionally lack experience handling certain types of food, guidance and training on how to actually execute food donations are also necessary. Along with the aforementioned issues, customers need to be made aware of the negative effects that food waste has on the environment and how much it contributes to carbon emissions. Additionally, people must be aware of the current resource limitations. The WRAP experts believe that local infrastructure, like separate food waste collections, can also have an impact on consumer behavior by raising awareness of the quantity of food waste and encouraging consumers to alter their food-purchasing, consumption, and disposal practices. Different kinds of philanthropic organizations engage in these kinds of activities. The food banks are one of them. In the US, the first banks opened in the late 1960s. Under many other names and with the same objective, the movement has spread throughout the world. Food banks gather food from various sources, store it in a warehouse, and then give it to needy families and individuals through neighborhood social welfare organizations. Community centers, soup kitchens, food pantries, and childcare centers are among the organizations that get assistance from food banks and food rescue programs. The majority of food banks in Europe are affiliated with the European Federation of Food Banks or EFB, which was founded in 1986 and promotes the growth of food banks throughout

Europe. 253 European food banks are currently a part of the organization. According to The European Food Banks Federation (FEBA), 240 food banks in 21 countries delivered 16,000 tonnes of food items that were meant to be thrown away in 2010 (Papamonioudis & Zabaniotou, 2022). In the same year, only 79 food banks in France gathered 98,780 tonnes of food, according to the annual report of the French Federation of Food Banks. The French food banks work together with a variety of players in the food supply chain, including the retail, food, and agricultural cooperatives sectors as well as with consumers. About 28% of the entire amount of food goods received by the banks in 2010 came from the retail industry.

Items that fail to fulfill marketing criteria due to aesthetic considerations are given to charitable organizations and others. It helps communities overcome food insecurity by donating excess food products from the food and beverage sector to nonprofits that assist underprivileged members of the community. However, food donation and subsequent redistribution as a strategy of reducing food waste face a variety of challenges because of various transportation, legal, and financial issues. Product liability is one of the biggest legal obstacles. The regulation states that potential food givers are liable for any 'health repercussions of food' that recipients may experience. As a result, one requirement for food donation is that potential food donors must be certain that staff members of food recovery programs or in charity organizations have received training in the proper handling and storage of donated goods. From an economic standpoint, it would not be practical for a farmer to contribute unharvested food by incurring the labor, logistics, and transportation costs. Additional financial challenges include the price of storing and packaging donated food, finding paid or volunteer labor, and training workers in proper food handling and preparation techniques. Nevertheless, despite these challenges, scientists think that food redistribution is one of the most effective methods for reducing food waste.

5.7 FOOD WASTE TREATMENT

Previously, lower in the waste hierarchy was where the regulatory measures for food waste treatment were concentrated. However, in more recent years, the situation has changed, and the EU has begun to pay more attention to the issue of food waste and has launched new programs as well as changes to the aforementioned legislation. For instance, the European Parliament urged the European Commission to promote the discounted sale of food that was approaching its expiration date in 2012 (van der Ploeg, Jingzhong, &

Schneider, 2012). However, an analysis of the findings revealed that it takes time for businesses to seize new opportunities and for entire industries to adapt to new regulations as a result of legislative changes like the gradual elimination of regulations or the loosening of regulatory restrictions on the aesthetic requirements for many fruits and vegetables. Therefore, by putting an emphasis on behavior modification and sector-based approaches in its guidelines for the establishment of food waste prevention programs, the EU targets the food manufacturing and processing, food retailing and distribution, food services, and household sectors. In general, rules and regulations might address the majority of the factors that contribute to the production of food waste. For instance, the introduction of policies or the supply of guidelines for dates that appear on packages, including the information required to interpret these dates, can reduce uncertainty about food dates. Because a single policy typically only affects one part of the food waste reduction process, such as the social, economic, or financial aspects, its successful implementation is directly dependent on other legislative changes. Economic incentives are one of the essential factors in food waste reduction methods since they are important and frequently taken into account when making decisions, irrespective of the size and complexity of a problem. Financial tools can be thought of as devices that encourage participants in the food supply chain to alter their behavior in relation to the quantity of food waste they produce. However, both the study of the current environment and the opinions of specialists mention a lack of these kinds of incentives. These instruments can generally be classified into two groups. The first type of financial stimulus is one that "forces" actors to incur greater expenditures; the second type of financial stimulus, in the case of reducing food waste, offers additional economic advantages.

By eliminating disposal costs and implementing better logistical and managerial techniques, reducing food waste may result in lower operating expenses and more income. A company's enhanced ethical/social reputation could result in increased revenue. Larger profit margins in the industry, which expand the alternatives for dealing with the waste, could be another financial incentive for the food industry to reduce its food waste. Because of financial incentives and legislation that supports it, the business climate in the United States is favorable to food redistribution. This is because it enhances the corporate image. Subsidies, tax credits that will promote innovations and R&D in the field, and funding of prevention programs with a focus on reducing food waste are further financial incentives for farmers. Customers may be encouraged to reduce food waste by such economic justifications as cost savings from improved food planning and purchasing. Informational and public awareness efforts may help disseminate it. Even though most participants in the food supply chain view expired food items as food waste, they nonetheless have some economic value. The amount and quality of material left in the waste stream as well as the expenses associated with its disposal would be significantly affected by keeping the food waste stream isolated from other types of bio-waste, such as garden waste. It will enable the optimization of the processing routes for each bio-waste component and assist in avoiding harmful materials in the composition of food waste. A higher-quality product can command a higher price and be used in a wider range of end markets. Separate collection also helps to increase recycling rates, lower greenhouse gas emissions, and lessen the negative environmental effects of landfilling. Measurement and separation of food waste will help increase employee awareness of the amount of food waste they are producing and encourage attempts to reduce this. Methods of treatment are determined by the types of food waste collection systems. For instance, composting facilities that accept both food waste and garden waste could find it challenging to regulate the nutritional ratios in the feedstock, which could have an impact on the quality of the final product. It's also critical to remember that the success of the separate food waste collection system directly depends on the public's willingness to participate in the initiative and their positive attitudes toward it.

Food waste disposal does not necessarily have to come after the point at which it cannot be used for human consumption (Garcia-Garcia et al., 2017). It still qualifies as a valuable resource due to the enormous potential it possesses as a source of energy, nutrition, or carbon. The global food and energy balance would benefit from such an alternative use of food waste as well. Microbes use the process of anaerobic digestion to break down organic matter without the use of oxygen. Two separate byproducts are produced by the process: digestate, a liquid or solid residue, and biogas. A gas can be burned to provide power and heat, further processed into biofuel, or fed into the gas grid. The combined annual capacity of more than 200 anaerobic digestion plants in Europe is 7,750 thousand tonnes. Germany, Spain, and France are currently in the top three positions among the European nations that use AD plants. Anaerobic digestion has a number of financial and environmental profits in addition to reducing the amount of waste transferred to landfills. In comparison to some other waste management technologies, the building of AD facilities can be relatively quick and economical. Digesting food waste through anaerobic digestion might generate extra revenue by avoiding the expense of dumping it in a landfill and by selling power, heat, and fertilizers instead. The idea of extraction rates only applies to processed goods, how much of the processed good in question was produced by processing the originating good, which is often a primary food. For instance, cheese is made from milk, soy oil is made from soybeans, and flour is made from wheat. The size of an extraction rate determines the extent of food losses, but it also accounts for a portion of food waste that can be calculated by assuming that the maximum value of an extraction rate among the countries under discussion is the maximum possible value constrained only by currently available technologies, with no mention of any other types of limitations. The difference between the maximum and actual values of the extraction rates existing in countries could be considered as food waste, so factors like the availability of the technical and financial resources in a country, in order to reach the maximum value of extraction rates, are ignored at this stage of calculations. It is also crucial to keep in mind that the data are unavailable for some nations. The outcomes of these computations, for instance, reveal that Sweden has the lowest rates of extraction for cereal goods, or the largest percentage of needless cereal waste.

Chapter

6

FOOD WASTE MANAGEMENT IN INDUSTRIES

Content

6.1 INTRODUCTION

These days, there is increased interest in environmental management. Understanding the significant environmental effects on the community is necessary before weighing the benefits and drawbacks of implementing various degrees of environmental management. The United Nations Economic Commission for Europe pledged to uphold the responsibilities brought about by the specified declaration and environmental accords. By assisting the UNEP and other international organizations in their work, they pledged to intensify their efforts toward sustainable development (Clark, 2007). Growing environmental awareness necessitates more complex industrial waste and emissions control, as well as quicker and more comprehensive on-board diagnosis. Pollution can come from both stationary and moving sources.

6.2 EMISSIONS

Similar to hazard analysis, there is growing interest in achieving zero emission in power plants. In 1999, there was more focus on the quantification, control, and modeling of wastewater discharges. Operations related to wastewater have been significantly impacted by the regulatory requirements introduced by the Clean Air Act Amendments. For cylinder-specific monitoring of petrol car engines, chemical sensors that operate at high temperatures and enough to detect the smallest time variations between an oxidizing and reducing environment can be utilized. When gas molecules, such as carbon monoxide, react with oxygen, the sensor signal is reduced, whereas when nitric oxide decomposes, oxygen is released on the surface and the sensor signal is increased. In 1975, the first study on silicon-based gas-sensitive devices with a palladium catalyst was released. New avenues for emission control were opened up by chemical sensors with catalytic metal functioning at high temperatures. They have a relatively simple working theory that relies on surface phenomena with very low time constants, allowing for extremely quick responses. They also represent a fusion of heterogeneous catalysis, material physics, and electronic technology. The Clean Air Act (CAA) mandates that industrial facilities handle the issue of unintentional releases of controlled substances to the public and establishes rules limiting wastewater-derived emissions of hazardous air pollutants. Water with a high volatile organic chemical content was stripped and released into the atmosphere in a publicly owned plant treatment in Virginia, which was managed by industrial users. In order to meet with state and federal clean air legislation, the standards for publicly owned installation treatment established a financially viable regulated technological approach. Closed sewage systems, nitrogen- or gas-blanketing vessels, and scrubbers on biological reactors are just a few examples of conventional treatment methods that can be employed to lessen the VOC (volatile organic compounds) content of waste and adhere to air emissions regulations. Many wetland systems are aimed at wastewater treatment, which is followed by auxiliary advantages like public use and wildlife habitat.

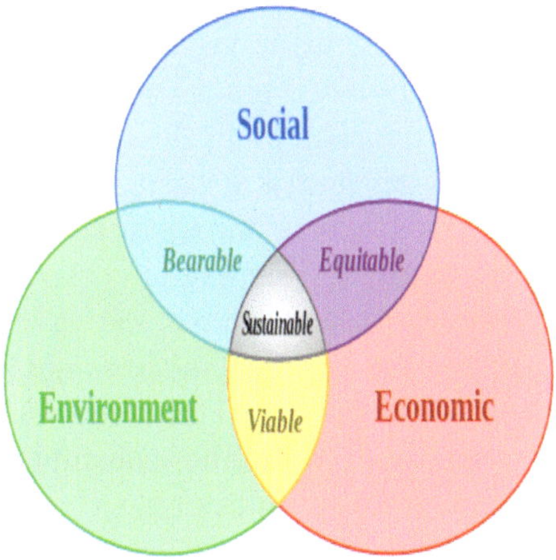

Figure 6.1. Sustainable development.

(Source: https://upload.wikimedia.org/wikipedia/commons/thumb/7/70/Sustainable_development.svg/512px-Sustainable_development.svg.png?20230308135220, Creative Commons License).

Sustainable development (Figure 6.1) could be used to create aquatic systems, more notably waste-stabilization ponds. In high-rate oxidation ponds, the spatiotemporal dynamics and removal effectiveness of pollution-indicator bacteria were investigated. Although the season had an effect on the removal of germs, retention duration was always well connected with removal. A fundamental barrier to the development of floating aquatic plant systems was revealed to be the high land area requirements and restricted cool season growth in temperate areas. Duckweed treatment systems are effective in removing algae, total suspended solids, and BOD, but their poor ability to remove nitrogen prevents their extensive use (Brix, 1993). The processes for water reclamation that are currently accessible to meet water-quality criteria have been analyzed and summarized in several articles and reviews. Gravel-bed hydroponic wetlands in Egypt were evaluated on a large scale, and the results showed that they were capable of eliminating parasite eggs and creating effluents that complied with WHO microbiological requirements for reuse. Traditional reclamation methods were deemed to be too expensive for Brazilian rural populations, thus treatment of degreased wastewater through grass was tested at locations with encouraging outcomes.

Traditionally, the dairy effluent has been utilized to dispose of a substantial amount of the chemicals used in the dairy sector. The kind and quantity of chemicals utilized, as well as the wastewater treatment technique used, had a significant impact on how polluted the environment was. One of the two instances related to waste reduction strategies was an ice cream business. BOD (Biochemical oxygen demand) was decreased by 63 percent once the proper procedures were put in place, and the investment expenses for the post-control wastewater treatment system fell by 43 percent. The ecological assessment of dairy chemicals demonstrates the environmental effects of certain compounds. In the UK, winter and summer milk samples were taken and examined for ortho- and non-ortho substituted polychlorinated biphenyls. For the ortho and non-ortho substituted polychlorinated biphenyls (PCBs), the determined values ranged from 0.05 to 0.6 ng/kg and 2 to 25 ng/kg, respectively. In the neighborhood of a municipal waste incinerator in Germany, the amounts of polychlorinated dibenzo dioxins and dibenzofurans in human blood and human milk from non-occupationally exposed persons were measured. It was determined that there should be no direct health risk associated with PCDD/F ([polychlorinated dibenzodioxins (PCDDs) and polychlorinated dibenzofurans (PCDFs)] emissions from the nearby waste incinerator because the individual amounts of PCDD/F in blood fat varied within safe ranges.

6.3 ADDITIVES FOR CLEANING

Surfactants seem to be more harmful than acids and alkalis, which will be replaced in the context of a cleaner environment. This does not, however, imply that cleaning should only be done with nitric acid and caustic soda. To maximize the detergent-capacity for cleaning, additives are added. The environment must be taken into account when picking additives. This causes a dilemma for the industry because every new product needs to have its environmental impact assessed. The end-users frequently purchase a mixed product and struggle with how to handle the chemicals it contains.

As a result, the producers are required to share their research on how their products affect the environment. For example, gas-liquid chromatography was used to identify dichlorodiphenyltrichloroethane (DDT) residues in multiple milk samples, and potential associations between DDT exposure and breast cancer were looked into. Although the measured DDT levels in human milk were partially attributable to DDT in fatty foods and partly to anti-malaria campaigns, no conclusive link between DDT and breast cancer could be made. When the dairy plant uses sources like surface water, groundwater, and condensates, it is important to pay close attention to maintaining quality standards. A dairy factory's quality management system should include these water quality controls, stating the frequency and metrics to be checked. Variable flow rates, seasonality, high organic loading, BOD and COD, low pH, and high carbon-to-nitrogen ratios are all characteristics of fermentation industry wastewater. Anaerobic biological wastewater treatment methods at a German brewery performed better than aerobic ones in studies that evaluated their effectiveness. The former treatment achieved 91 percent COD reduction, while the aerobic treatment averaged 76 percent reduction. 85 percent of the COD was removed from brewery effluent in an anaerobic fluidized bed reactor operating at 25 °C, and the hydraulic retention duration was 2.5 h. An average amount of biogas was produced each day at a rate of 0.45 m^3 kg COD eliminated, of which 72% was CH_4. The pH of brewery wastewater increased from 6.5 to 8.9 after seven days of aerobic treatment, while the BOD and suspended particles decreased by 67 and 75%, respectively (Yang, Yang, Zhang, & Lv, 2005).

Using laboratory columns, it was determined how methane generation and landfill leachate were affected by different rates of mixing brewery wastewater with municipal solid waste. It was discovered that the co-disposal of brewery wastewater increased methane production while having no detrimental effects on the pH or volatile fatty acid concentrations of the leachate. However, at very high organic loading rates, leachate quality briefly declined as higher-than-normal amounts of acetate and propionate were found. The findings of a study that considered the possibility of organic nitrogen molecules, such as those found in high concentrations in brewery effluent, interfering with aqueous chlorine disinfection showed that wastewater with high ratios of total organic nitrogen to ammonia performed poorly. In Germany, the operation of anaerobic wastewater treatment facilities, especially those for brewery waste waters, was examined. In activated sludge plants, a survey was done to determine whether filamentous bacteria were present. In sludge from a facility handling brewery wastewater, a gram-positive, Neisser negative filament resembling both *Microthrix* spp. and *Nostocoida limicola* was the predominant filament.

Both laboratory and an industrial-scale equipment were used to test the treatment of winery effluent. A full-scale, modular, multistage activated sludge treatment facility was investigated for the treatment of winery wastewater. When the influent COD (Chemical Oxygen Demand) ranged between 2000 and 9000 mg/l, the elimination of COD was on average 98 percent. The plant was shown to have a low running cost and to be adaptable enough to handle significant fluctuations in temperature, hydraulics, and COD loading rate. Studies conducted in Portugal describe a 1 m^3 pilot plant's capacity to treat winery wastewater using an adsorption activated sludge technique that was designed for use. In order to achieve high removal efficiencies from the wastewater of 96.0 and 97.7 percent, COD and BOD5, respectively, it was necessary to promote biomass buildup and stabilize the reactors through the addition of nitrogen and phosphorus fertilizers. Several attempts have been made in France to increase the efficiency of aerobic winery wastewater treatment by the seeding of winery bacteria, pH correction, and addition of nitrogen and phosphorus nutrients. The treated effluent had to be safely discharged to the municipal treatment system by reaching sufficient treatment levels in a short period of time. Using Fe^{2+}, Cu^{2+}, Zn^{2+}, and Co^{2+} ions and metal oxide catalysts in suspension, the catalytic oxidation of p-coumaric acid, a molecule that serves as a sample of the polyphenolic fraction normally found in wine distilleries was studied. The distribution of intermediate products was shown to be influenced by solution pH while the rate of p-coumaric acid breakdown was boosted by catalysts. Vinasse is a noteworthy issue in the treatment of wastewater. It was shown that minimum HRTs ranging from 4 to 8 days were required for the 90 percent COD elimination of vinasse purified by mesophilic aerobic/mesophilic anaerobic or thermophilic anaerobic treatment, respectively. Except for when the hydraulic retention time (HRTs) fell to their lowest values, a substrate usage model proved accurate. In tests that simulated the three main phases of carcinogenesis, resveratrol, a phytoalexin found in grapes and other foods, was isolated from grape skin waste and demonstrated to have cancer-chemopreventive potential. These findings imply that resveratrol, a staple of the human diet, should be researched for its ability to prevent cancer in people.

Continuous-stirred tank reactors were utilized in comparison studies of the anaerobic digestion of untreated fermented molasses. The observed variation in the digestion process may be due to increased quantities of phenolic chemicals in the untreated wastewater interfering with and inhibiting anaerobic bacterial activity. A review of several alcohol distillery plants that were treating sugar beet molasses revealed that the use of open stream and recycling of spent mash had an impact on wastewater volume by up to 25%. The COD load remained essentially unchanged with recycling, whereas total dissolved solids increased significantly(Demirel & Yenigun, 2004). Wastewater from a sugar beet molasses distillery included twelve potentially inhibitive phenolic chemicals, which were separated and named as follows: six benzoic acids, two cinnamic acids, three phenolic aldehydes, and one phenolic alcohol. Recently, it was proposed to use a combination of several media, such as black liquors, lime, and polyacrylamide, to precipitate and flocculate the wastewater from the fermentation of ethanol. This approach produced water that is acceptable for animal feeds and microbiological culture media as well as water that may be reused in the production of ethanol. Technology has been created to recycle or remediate by-products of the sugar-processing industry. Full-scale upflow anaerobic sludge blanket reactors were used in studies on the anaerobic treatment of various distillery effluents, and the treatment of effluents from sugar cane-based distilleries with COD concentrations of 60,000 to 160,000 mg COD/l was highlighted (Boudouropoulos & Arvanitoyannis, 2000). Numerous soft drink bottling factories that use anaerobic filters for pretreatment have reported issues with odor, biological reactor issues, and problems that have affected on-site plant operations. Despite the high degree of COD removal offered by the processes, changes in hydraulic and organic loading, contamination by petroleum oils, and the transfer of solids from the clarifier to the biological system all hindered the system's efficiency. High-pressure reverse osmosis was chosen over the other membrane filtration technologies in a recent pretreatment technology study for a soft drink company. After doing a cost analysis on the pilot testing of several membrane filtration systems, biological pretreatment was rejected. Due to its high processing capacity and 99 percent BOD removal, this technique was chosen.

6.4 AGRICULTURAL WASTE

Waste from the agriculture sector turns out to be a huge source of materials with a natural origin. Utilizing this waste yields natural goods with low-cost raw materials, such as antioxidants. A good example of a Greek company conducting research on citrus peel after juice extraction and pressed grape skins from winery waste is VIORYL SA. The primary characteristics of this method include solvent selection for effective extraction and recycling, environmental safety, and choosing conditions for non-destructive product recovery. A thorough literature search was done on the use and disposal of solid vegetable, fruit, and other organic waste. Orange, grapefruit, and bergamot peels were used to extract common flavonoids with pharmacological characteristics as hesperidin and naringin. By extracting the peel using hot water and calcium, both flavonoids were made available. Hesperidin and naringin yields increased as a result of peel maturity and recycling of the extracting liquor. Hesperidin and naringin had the highest yields of 15.2 g/kg and 12 g/2 kg peel, respectively. In order to research the variations that occur at these temperatures, sequencing batch reactor pilot plants for treating wastewaters from potato processing were operated at varied temperatures. Nitrification was significantly decreased at low temperatures, despite the fact that there were only minor variations in effluent soluble COD concentration that were related to temperature. At the highest temperature, there was a considerable decrease in phosphorus elimination. Insecticide water remediation using waste mushroom compost as an adsorption medium and a source of microorganisms was also explored. It has been demonstrated that the compost can absorb carbamate insecticides from aqueous solutions and doesn't hinder growth in the concentration range of 0 to 30 mg/l. Another pertinent study revealed that composting potato peel waste for 3 to 4 weeks led to a weight loss of 51% and a dry matter loss of 27%. The garbage was supplemented with sawdust and urea. For the treatment of wastewater from the production of olive oil, the effectiveness of an anaerobic upflow blanket-type reactor was assessed.

The handling of these materials was covered in various studies. Dissolved air flotation wastewater treatment systems at poultry processing factories produce biosolids, which pose a substantial disposal issue. The effectiveness of utilizing lactic acid to stabilize offal and dissolved air flotation sludge was assessed. In order to evaluate the quality of the sludge, studies concentrated on measuring pH, ammonia, protein, fat, and bacterial pathogens. An extensive analysis of the foaming and thickening issues in the activated sludge system for wastewater from chicken processing was done. Process type, food microorganism ratio, dissolved oxygen levels, liquid operating temperature, changes in production rate, and aeration cell inlet/outlet design were some of the parameters that were looked into. Recently, the origin of odors from numerous previously addressed control methods, such as combustion in boilers, wet scrubbers, and biofilters, was studied. Since the 1970s, pesticide and herbicide-related water pollution has been acknowledged as a very serious issue (Stalikas & Konidari, 2001). Many of the people had exposure to industrial chemicals and product contaminants like polychlorinated biphenyls, polychlorinated dibenzo-p-dioxins, and polychlorinated dibenzofurans. Similar to how it was with DDT and other chlorinated insecticides, tested persons had high background levels of PCBs, PCCDs, and PCDFs. These lipophilic substances, which are frequently regarded as potential carcinogens and are prevalent in fatty foods, could easily enter humans if they are prevalent in the terrestrial environment. The amount of pesticides that enter rivers and lakes is a major concern because if pollution gets into drinking water, it could affect aquatic life and people. The EC Drinking Water Directive, which mandates a maximum concentration of any pesticide in drinkable waters of 0-1 g/l, addresses these concerns. Despite the fact that the toxicity of different pesticides to humans varies widely, epidemiological data on the impact of pesticides on human health are generally lacking.

6.5 RESEARCH FINDINGS

Although research is beginning to indicate that excellent management methods can reduce losses, secondary economic costs, particularly those related to the water business, are still substantial. Numerous studies have looked into herbicide contamination of shallow groundwater beneath claypan soils. The effects of atrazine and arachlor application rates were found to be smaller than the effects of spatial variability. In alluvial aquifers in the USA, no appreciable variations in arachlor, atrazine, or nitrate concentrations were found between pre- and post-flood distributions. Low amounts of H_2S and VOCs were simultaneously eliminated by granular activated carbon biofilters operating at a wastewater treatment facility and compost made from agricultural and other waste. The presence of more gram-negative bacteria and varied degrees of environmental stress were seen in the microorganisms in the media. The elimination of the compounds, which was 99% for the H_2S and 70% for the oxygenated and aromatic hydrocarbons, was not significantly affected by lowering pH. To evaluate the decomposition of chlorophenol by different inoculants, straw compost and bioremediated soil, with that of native soil bacteria, a pilot-scale composting of chlorophenol-contaminated soil was set up. The biodegradation was discovered to be extremely effective, quick, and independent of the type of inocula.

Researchers investigated the generation of activated carbon from agricultural waste, such as grape seeds, apricot and cherry pits, and almond shells. These studies used a low temperature technique, and it was shown that the structure and composition of the feedstock had an impact on the characteristics of the activated carbon. The nitrification and denitrification of wastewater from settling manure in an activated sludge system with the addition of molasses as an additional carbon source was another experimental issue. With the only carbon source being the supernatant from manure, complete denitrification proved to be impossible. Only trace levels of nitrogen oxide were discovered when nitrifying and denitrifying conditions were enhanced. Due to its high consumption rate, olive oil manufacturing is crucial, especially for Mediterranean nations. Frequent washing occurs while milling olive paste during the manufacturing of olive oil, producing huge amounts of wastewater that are wasted. Wastewater extracts have strong antioxidant effects because they include natural antioxidants. The latter would be a promising inexpensive source of antioxidants if it were effectively collected and could demonstrate that the waste from the production of olive oil has a potent anticarcinogenic activity. Numerous papers discussed the impacts of a wide range of pollutant types, including specific complex compounds and metals, on aquatic species in freshwater and seawater. Cell culture plates were made for a quick bioassay to test the specific toxicity of herbicides and other phytotoxins against cyanobacteria. Green algae and cyanobacteria were represented by *Oscillatoria cf. chalybea* and *Selenastrum capricornutum*, respectively (Schrader, de Regt, Tidwell, Tucker, & Duke, 1998).

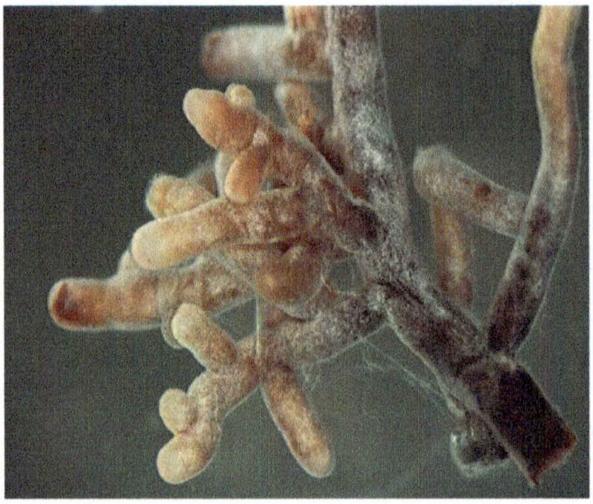

Figure 6.2. Ectomycorrhizal fungi.

(Source: https://upload.wikimedia.org/wikipedia/commons/a/a5/Mycorrhizal_root_tips_%28amanita%29.jpg?20051213213118, Creative Commons License).

6.6 SENSITIVITY

The sensitivity of various ectomycorrhizal fungi (Figure 6.2) to Cd^{2+}, Pb^{2+}, Zn^{2+}, and Sb^{3+} was examined in locations free of harmful metal contamination. While Pb^{2+} and Sb^{3+} individually had minimal impact, interspecific heterogeneity in sensitivity to Cd^{2+} and Zn^{2+} was shown to be significant. The toxicity of Cd^{2+} and Zn^{2+} was, however, occasionally reduced by the presence of Pb^{2+} and Sb^{3+} in these media. Mercury, cadmium, and cobalt were found to be the most harmful heavy metals to specific strains of wood-rotting basidiomycetes. Only one strain of iron oxidizing bacteria, *Thiobacillus ferrooxidans*, thrived on Fe^{2-}medium with 1.25 mM of sodium molybdate. Additional research identified the molybdenum-inhibitory location for the iron-oxidase enzyme system and the molybdenum resistance mechanism in Funis 2-1 cells. There were no discernible patterns in the numbers of taxa among the study sites and dates, according to faunal species sampling conducted close to a lead smelter in South Australia. Further, *Heteromastus filiformis*, *Capitella capitata*, *Glycera americana*, and the crab *Tanais dulongii* were really found in large numbers at intertidal areas with substantial metal concentrations in the sediments. There is a lack of data demonstrating that earthworm populations were negatively impacted by chlorinated chemicals found in the environment. In laboratory tests, DDT or dieldrin doses as high as 64 ppm had no effect on the life of *L. terrestris* and *A. caliginosa*. Applications of DDT at rates surpassing 100 ppm also had no effect on earthworm survival (Fries, 1991). The most heavily 2,3,7,8-tetrachlorodibenzo-p-dioxin (TCDD)-contaminated area was found to have quite typical earthworm activity, however, there may have been a difference in the ratios of the most abundant species, *A. caliginosa* to *A. rosea*.

The most notable examples of the harmful impacts of organic pollutants on terrestrial wildlife have been supplied by birds. In lab experiments, the effects of chlorinated hydrocarbons on eggshell thickness were further supported. Aquatic predators experienced reproductive problems the most frequently and severely, followed by terrestrial predators, insectivores, and herbivores. Cattle and horses have been severely poisoned by PCBs and PCDD due to improper disposal of polluted waste oils. Dioxins, which are carcinogenic, have gotten into the food supply and quickly destroyed the established system for ensuring food safety. It is thought that the European Union was aware of at least some of the current safety issues but held off on enforcing stronger regulations due to the increased production costs that would have resulted. However, exceptionally high hormone traces had been found in beef carcasses coming from the USA before dioxin was discovered. Compared to colonies in non-oiled locations, kittiwakes from a colony in the Shetland Islands polluted by the crude oil spill had comparatively strong breeding success and foraging activity. Ichthyoplankton surveys show any consistent variations in taxa or fish loads due to treatment either at the surface or at a depth of 20 m because of the high levels of spatial and temporal variability. However, even at a 0.1 percent dilution, a mixture of primary and secondary wastewater sludge had a detrimental impact on the health of herring embryos, the survival of hatchlings, post-hatch larvae, and larval nutrition in a lab setting. Only the release site was affected by toxins. It has been demonstrated that chlorine, which is frequently used to disinfect wastewater and power plant discharges, is hazardous to anchovy eggs and larvae at concentrations much below the dosages advised for treatment.

Food items are greatly impacted by the frequency and mechanism of organic compound introduction into the environment. The primary causes of low-level dioxin and furan environmental pollution have been identified as combustion sources, such as municipal waste incinerators and vehicles. The different natural and artificial sources of nitrate pollution of groundwater, the sources of nitrate exposure for people, the potential health implications, and the difficulty in managing nitrate protection for groundwater were all described by researchers. In a case-control study, the association between nitrate levels in private and public sources of drinking water and stomach cancer mortality in Wisconsin was examined. Exposure to nitrate in drinking water sources could not be significantly linked to deaths from stomach cancer. However, there is still an opportunity for development. New technology in the glass industry has reduced the gas discharge of pollutants like NO_x and particles as well as characteristics like BOD. While achieving zero pollution is essentially unachievable, improving the environment is a continuing objective. Industry offers a variety of methods and pollution-arresting programmes that can be used to satisfy present-day or even upcoming legal requirements. In order to lessen their negative environmental consequences, to ensure worker health and safety, and generally to reduce individual consumption, the usage of dangerous heavy metals is subject to tight regulations.

A waste management strategy for the chemical and petrochemical sector comprised pretreatment facilities, stringent regulation of landfill content, and regional management of hazardous wastes in Brazil (Ferraz de Campos et al., 2021). Three steps were taken to improve the groundwater quality at a significant industrial complex: baseline data collection, the creation of monitoring programs, and suggestions for novel management techniques. The main reason for this facility's waste management/minimization program was an increase in contaminated areas brought on by subpar landfill operations at a chemical business in the Netherlands. To adhere to the restrictions placed on phenolic compounds transported for biological waste treatment, a recovery procedure for nitrophenols that takes into account potential thermal instabilities was devised. For the purpose of eliminating dissolved organics from refinery effluent, biological treatment is a crucial technique. To compare the capacity to biotreat spent sulfidic caustic from a refinery, microbial cultures were investigated. Over the past few years, anaerobic treatment of petrochemical effluent has gained a lot of attention. A pilot investigation of the upflow anaerobic sludge blanket reactor (UASB) process for treating wastewater from the production of purified terephthalic acid was carried out. P-toluic acid, one of the main components of purified terephthalic acid (PTA) effluent, hindered biodegradation. Phenolic chemicals are prevalent in both industrial and municipal streams as well as natural products. In particular, if they are chlorinated, which produces chlorinated phenolic compounds, which are very poisonous even at ppb levels, their presence may prove to be hazardous. Despite their poor environmental performance, the phenolic chemicals found in this effluent are employed as a feedstock for the creation of polymers, medicines, and colors. Therefore, liquid-solid phase extraction is used to extract phenols from water samples.

Many current emission-control techniques are currently ineffective due to strict emission regulations. In response to the Clean Air Act of 1970, the impact of automotive catalysts to counteract incomplete combustion in automobile engines was examined. Studies of the emissions of unburned hydrocarbons and carbon monoxide at the upstream and downstream sides of a catalytic converter revealed that these emissions might be significantly reduced by using the right ignition-control techniques. Utilizing the energy contained in fossil fuels has been the fundamental driver of economic expansion in contemporary industrialized society. Fossil fuel consumption has become a tantalizing force despite the fact that humans have learned to exploit them fairly effectively. This is because of emissions, spills, leaks, strip mining, and carbon dioxide accumulation that result in highly dangerous and threatening conditions such as global warming, rising sea levels, and earthquakes. In the USA, public concern about the environment has never been higher by 1990. After a drop in environmental concern in the 1970s, there has been a large and sustained rise in public knowledge of environmental problems. The usefulness of preserving the services given by environmental systems to humans is a major topic of discussion in contemporary environmental policy. Rejecting anthropocentric and economic considerations directly is a common way to express care for the environment. Only once the causes of the pollution have been correctly recognized can the pollution abatement schemes be put into action. These expenses can occasionally be attributed to incorrect specifications or the inapplicability of property rights to the use of shared resources. Alternatively, the presence of technological non-convexities prevents the creation of competitive markets for pollutant emission rights when well-defined property rights are present and their enforceability is achievable.

6.7 ENVIRONMENTAL DEGRADATION

According to current trends in environmental degradation, the planet will become uninhabitable if the pace of environmental deterioration keeps increasing. Reports on ozone layer loss and global warming indicate that immediate action is required to avert an impending disaster. Sunlight provides essential energy to all living things. Plants use it to transform carbon dioxide and water into intricate, high-energy compounds. Photosynthetic organisms are able to absorb and use more than 10^{17}kJ of solar energy (Hungate, 1982). During the NASA space missions of the 1960s, solar electric power was created to solve a specific power issue. Solar radiation, like many other space technologies, was initially used in isolated locations where power was needed yet impossible to obtain through other means. The possible financial benefits from producing solar power rely on the operator's cost for purchasing exported solar power as well as the cost of national power, which varies widely. There are additional infrastructure and participant familiarity barriers to overcome because solar energy produces direct current or DC electricity that must be converted to alternating current in order to interface with the grid. As more experience is gained, these are being quickly addressed. Solar power is still a rather expensive technology, but the cost is steadily going down and the market is growing. New, less expensive technologies are now being developed, and the World Energy Council anticipates significantly higher sales by 2025. The prospects for this technology are very bright, especially as concerns about climate change and environmental degradation move up the political agenda. The world has produced over 700 MW of wind energy, making it the energy technology with the fastest growth. The oil crisis of the 1970s is one of the factors behind this sudden increase in wind energy, but more recent drivers include the pressing demand for electricity in developing nations and the quest for non-polluting energy sources.

Because it is technically sound, has widespread backing, abundant resources, and is quickly approaching cost parity with traditional sources of electricity, wind energy is currently a practical choice for producing electricity. Future cost reductions and performance enhancements have high possibilities. Further research on dynamics is reflected in economic optimization, which in turn depends on an accurate understanding of the complicated nature of wind turbine loadings. As a result of global warming and the greenhouse effect, it has been anticipated that ice caps will melt and climates will change in the future. This disruption in the Earth's ecosystem is largely related to the emissions of greenhouse gases into the atmosphere and is raising serious concerns. The greenhouse effect, brought on by increases in atmospheric CO_2 levels, is generally acknowledged to produce substantial warming of the Earth's surface and potentially catastrophic environmental effects. The four basic methods for capturing carbon dioxide are membrane systems, cryogenics, adsorption, and absorption. Molecular sieves are frequently used in gas adsorption, although they call for the release of absorbed gas after it has been trapped.

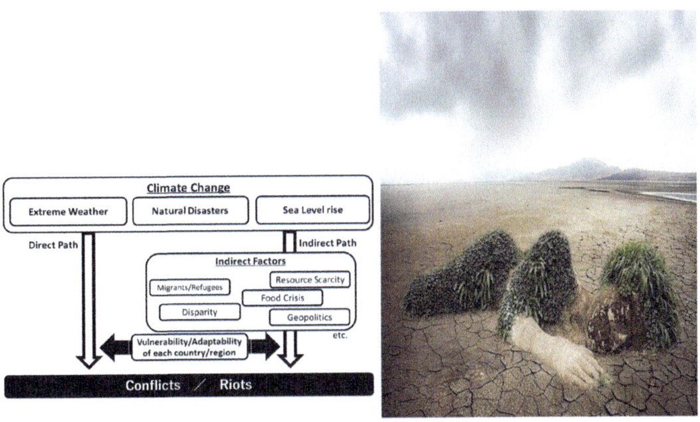

Figure 6.3. Climate change.

(Source: Sekiyama T. Climate Security and Its Implications for East Asia. Climate. 2022; 10(7):104. https://doi.org/10.3390/cli10070104, Creative Commons License).

It is preferable to change the procedure for releasing the gas rather than the temperature because the absorber can start working again much more quickly. This technique is unfavorable for CO_2 capture from the traditional generation process due to its low capacity and poor sensitivity. Each type of power plant can use a different solvent to achieve physical and chemical absorption. The ability to produce liquid CO_2 that is ready for pipeline transit is a benefit of cryogenic operations. Although they are utilized commercially, membranes need to be improved and their price must decrease before they can be more commonly employed for CO_2 capture.

6.8 CLIMATE CHANGE

Consideration should be given to scientific solutions to climate change (Figure 6.3) as a means of reducing greenhouse gas emissions while achieving human goals for a higher standard of living. Capturing and storing CO_2 is assumed if CO_2 emissions must be significantly decreased (Möllersten, Yan, & R. Moreira, 2003). Science has only recently been interested in these technologies, yet there are currently working storage methods all across the world. A clean, reliable, durable, long-lasting, and renewable energy source is hydropower. It is by far the most productive renewable source, contributing close to 5% of global electricity production. The three most widely used renewable energy sources wind, biomass energy, and hydropower account for more than 90% of all electricity production. Based on hydropower's potential and efforts to reduce the greenhouse gas emissions from power plants, which contribute to global warming, it is likely that hydropower will play a role in fostering long-term, sustainable global economic growth. In many places across the world, geothermal energy i.e., heat from the Earth can be used to partially replace conventional fuels. With the expansion of power generating stations, heating systems, food processing facilities, and greenhouses around the world, the great potential of this clean, dependable energy source is only now beginning to be realized. Geothermal energy is anticipated to assist in meeting our needs in the next few years as ecologically friendly energy sources become progressively more crucial. Although it is not the only answer to the world's need for clean energy, it makes a substantial contribution and has a lot of room for growth in the future. Humanity will soon be able to utilize the tremendous geothermal heat stored in rocks at any depth, not just in volcanic zones, due to advancements in science and technology. Geothermal energy may very possibly meet up to 25 to 50 percent of the world's electricity and heating needs when it reaches this degree of development (Ravi Kumar, Krishna Chaitanya, & Sendhil Kumar, 2021).

An electrochemical device identified as a fuel cell uses an electrode-electrolyte system to continually transform chemical reactions and oxidant fuel energy into electrical energy. High efficiency and emission levels below those required by the highest environmental requirements are characteristics of fuel cells. According to their operating temperature and pressure, fuel cell systems can be categorized as high-, medium-, or low-temperature/pressure systems. Based on the types of used fuels and oxidants, gaseous, liquid, and solid, another classification may be made. The initial high cost of catalysts, membranes, and accessories, the weight of the fuel, electrode degradation in conjunction with total carbon dioxide removal, and infrastructure gaps for the distribution and transportation of hydrogen are some of the drawbacks. Even though it may seem unusual, one of the safest methods for cleaning up contaminated land is to harvest garbage. Land that has been polluted with metals could be revitalized affordably and sustainably by using plants that selectively absorb toxic metals from the soil and then just collecting and discarding them. Prior to the deployment of bioremediation and phyto-remediation on a broad scale to detoxify metal-polluted soils, several practical and research hurdles still need to be overcome. However, this green strategy using hyperaccumulator plants to clean up polluted regions is thought to offer a lot of promise in the fight to preserve soil fertility, one of the most priceless natural resources in existence. The goal of the cost-benefit analysis approach is to choose initiatives and policies that make the best use of available resources. If advantages are considered to exceed disadvantages, the suggested approach could result in an improvement in social welfare. Although CBA is widely used and accepted, it has also come under harsh criticism from areas other than economics, including the social and environmental realms.

One of the main reasons the traditional cost-benefit analysis is not thought to be very useful to support environmental policy is the degree of uncertainty around the estimates of the external environmental costs. The European Directive on the promotion of cogeneration is very favorable to micro-cogeneration and small combined heat and power (CHP) plants on the basis of their assumed environmental benefits, and the European Commission is actively promoting distributed combined heat and power generation as a means of reducing greenhouse gas emissions. Predicting the effects of a planned action, typically a project or policy, on the environment before that project or policy is undertaken is called environmental impact assessment. Any goal of the Environment Impact Assessment (EIA) is to support the development decision; it is not an aim in and of itself. A quality

EIA guarantees that decision-makers have access to the best information possible when evaluating projects. The second layer of the EIA is concerned with its methodology and organizational structure, while the third layer is focused on the actual operations that take place inside its framework. The potential for the EIA process to serve as a platform for negotiations between the developer, public interest groups, and the planning authority is not always utilized. The resulting changes in environmental parameters through time and space as compared to what would have occurred without the project are the environmental consequences of a project.

Any of the previously mentioned environmental receptors, such as noise, air pollution, water quality, local unemployment rates, and crime rates, may be used as parameters. Physical and socioeconomic impacts can be direct or indirect, short- or long-term, local or strategic, negative or positive, reversible or irreversible, quantitative or qualitative, actual or perceived. Identification of prospective changes in these environment receptor markers is necessary for prediction.

The broad effect categories in regard to the project under consideration will have been determined through scoping. Regardless of the project or activity, if a certain environmental indicator showed a problem that was steadily getting worse in a given location, this should be taken as the baseline going forward. Although there are numerous ways to categorize prediction methods and models, they are not exclusive. All methods, in terms of scope, only cover a portion of the effects; however, some attempt to be more comprehensive than others. Partial techniques can be categorized based on the project type and the impact type. Others may be more normative, while some may be extrapolative. Normative techniques evaluate whether a project, in its environmental context, is adequate to accomplish intended goals by working backwards from desired outcomes.

6.9 ENVIRONMENTAL IMPACT STATEMENT OR EIS

A document that describes a proposed development or activity and discloses the possible, probable, or certain effects of that proposal on the environment is known as an environmental impact statement or EIS. Depending on the country, this document is also referred to as an environmental effects statement, an environmental statement, or an environmental impact assessment. An EIS should be thorough in how it approaches the topic, objective in how it approaches it, and sufficiently explicit to allow a fairly reasonable mind to analyze the potential environmental consequences, both positive and negative of implementing, or not implementing, that proposal (Doelle & Sinclair, 2006).

The following are typically included in an EIS: a thorough description of the proposed project or activity; a statement of the proposal's objectives; an adequate description of the existing environment likely to be impacted by the proposal; the identification of the likely environmental interactions between the proposal and the environment; the justification of the proposal; economic, social, and environmental considerations; and the measures to be taken with the proposal. These unplanned expenses, which make up around 10% of the total project expenditures, might be significant. When dealing with a sizable electrical generating plant and transmission network, they are very difficult. Pollution control expenses typically represent 20% of overall plant and equipment expenditures for the iron and steel industry, 12% for non-ferrous metals, and 11% for electricity-producing facilities. The audit is one of the best tools for enhancing the procedure. Its main goal is to confirm that the development was carried out in accordance with the restrictions imposed by the initial EIA procedure and the planning approval that went along with it. Monitoring adherence to the agreed-upon conditions, reviewing anticipated environmental impacts, modifying the activity or developing mitigation measures in case of negative effects, and evaluating the efficacy of environmental management are all tasks carried out by UNECE task forces. The climate in Europe and the world has changed noticeably in recent decades. Temperatures are rising, precipitation patterns are altering over most of Europe, and some places are experiencing an increase in the frequency of weather extremes. The majority of the warming seen over the previous 50 years is attributed to human activity, particularly the release of greenhouse gases, according to the UN Intergovernmental Panel on Climate Change (Dai, 2011).

6.10 AGRICULTURE AND CLIMATE CHANGE

Agriculture is impacted by climate change in different ways. Early sowing dates, improved crop growth, and a potential rise in crop output may be made possible by rising temperatures and increasing atmospheric CO_2. Extreme weather occurrences harm infrastructure, industry, and individual residences. Since 1980, many of the catastrophes that have occurred in Europe have been caused by weather and climatic extremes, such as droughts and heat waves. One of the major contributors to environmental pollution is the burning of large quantities of fuel in various types of furnaces. The accompanying ashes damage land and water, while flue gases pollute the atmosphere. One common indicator would be useful to compare the environmental impact of different fuels burned in varied facilities. A report defines a coefficient that indicates how damaging the combustion processes of a specific fuel in a specific installation are. This coefficient accounts for the fuel's composition, the installation's thermal efficiency, the type and process by which harmful compounds are produced in the combustion chamber, the effectiveness of cleaning devices, the emitter's capacity to disperse pollutants into the atmosphere, as well as the toxicity of various components. All of these elements are necessary for a dependable evaluation and comparison of fuel and installations. Both a total coefficient and an individual coefficient of harmfulness for each component found in the fuel gases may be computed. Each component produced during the combustion processes might have a variety of negative effects on the environment. It might have a big impact on things like the corrosion process going on in many different installations, as well as the health of people, plants, and animals. It must be remembered that the outcome of a combustion process could manifest itself in a variety of locations, sometimes far from the site of the activity itself. All of these findings need to be examined and factored into the inclusive measure of a compound's harmfulness.

The development of intensive agriculture in the industrialized world is under increasing attack because, despite being profitable in the short term, it has significantly reduced the biological complexity of farming systems through the imposition of strict production controls meant to boost energy and capital inputs. The voluntary acceptance of numerous sustainability-related rules and charters demonstrates the importance of the sustainable development concept to the growing global interest in environmental management. Most nations have now put into effect new legislation requiring an Environmental Management System (EMS). Environmental management has faced difficulty in determining which systems to create and how to apply them efficiently. The responsibility facing the environmental manager might easily become too much to handle once one takes into account the extensive management systems that are currently necessary for ISO 14001 certification. The first two standards to be made public are ISO 14001 and ISO 14004. Since October 1996, more documents pertaining to auditing concepts and practices, auditor credentials, and life cycle assessment principles have been approved and released. Many ISO 14000 committee documents have progressed to the international standard level since 1997 (Pun, Chin, & Lau, 1999). These documents concentrate on life cycle assessment, environmental management terminology, and environmental labeling and declarations. Environmental management and standardization have been combined through the use of several suggested standards. In an effort to make the proposed standards as practical, user-friendly, and acceptable to all the nations that have adopted ISO 14000, numerous organizations and individuals from all over the world have worked on them. They are environmental goals and broad objectives that the company supports. The established environmental policy must adhere to the relevant environmental laws. The evaluation of environmental consequences is a requirement for choosing objectives and aims in the right way. Although ISO 14000 does not mandate that all of these issues be fixed at once, it is a guiding concept in selecting and prioritizing objectives and targets that should address substantial environmental effects. The creation of an environmental policy that demonstrates the commitment of top management of the organization to compliance with existing laws and ongoing development is part of the International Standardization Organization's definition of an EMS. Based on this policy, objectives and targets are determined, and a program for environmental management is developed to carry them out through organizational structures, management controls, and accountability. Along with these tasks, it also entails reviewing and improving the management system, addressing issues, and measuring and monitoring progress toward the goals. An EMS does not, however, provide precise performance objectives.

The standard also mandates that the organization make sure that its environmental policy complies with two key requirements: the policy must be appropriate to the environmental impacts of its activities; and a commitment must be stated with regard to continuous improvement, pollution prevention, and compliance with relevant environmental legislation. The company's environmental policy must be outlined, put into practice, kept up to date, shared with all personnel, and made publicly available. Establishing and maintaining an EMS that includes all standards as outlined in the standard is one of the most fundamental requirements of an EMS. The trade-off between costs and advantages will ultimately determine whether or not ISO 14000 standards are adopted. It is expected that having uniform standards will reduce trade barriers and misunderstandings. It is generally accepted that a company's EMS certification by an internationally renowned registrar may strengthen the effect of the standards with outside groups and governments. Many companies that process hydrocarbons debate whether or not to devote their effort and money to obtaining ISO 14000 accreditation. Like ISO 9000, ISO 14000 is a procedure for confirming that the operating company is making an effort to meet its environmental goals and policies. According to the criteria, businesses should create an emergency management system as the foundation for their individualized policies and procedures. Although implementing EMS won't ensure that the company's environmental performance will improve, it can greatly increase its chances. Along with the marketing benefits, EMS can establish an environmental culture and persuade workers to embrace a work ethic that could lead to increased economic performance (EP).

Because the standard was developed to be relevant to all businesses regardless of kind and size in a variety of geographical, cultural, social, and economic contexts, the answer to the question of whether ISO 14001 creates EP requirements is no. In reality, it aims to strike a balance between corporate and socioeconomic needs while promoting pollution control. ISO 14001 has requirements that should lead to a higher and continually growing performance level even though it does not establish definite EP requirements. Investments in software, labor, equipment installation, and perhaps third-party auditing are all necessary for ISO 14000 certification. The adoption of these ISO standards and guidelines promoted improved forestry practices all around the world (Klooster, 2010). The management of forests entails a very complicated web of economic, social, and environmental challenges. The ISO methods and standards contributed to a better understanding of these intricate global problems and provided a workable framework for their local application. Trade and business representatives from the Netherlands have consistently offered specific recommendations for putting into place an ISO-14000-based system of certification and have backed Canadian efforts in this regard at international forums.

The implication of creating an environmental strategy that supports sustainable development and incorporates stakeholders' challenges, concerns, and requirements is being understood by business leaders around the world more and more. To implement stakeholder management procedures and support a sustainable global economy, the ISO 14001 standard for EMS provides a collection of useful and certifiable tools. Having access to and using correct environmental data may be necessary for a bank to be able to modify its loan and investment operations to take environmental considerations into account. With the introduction of ISO 14000, a chance to increase the knowledge base on the environmental performance of industrial organizations may present itself. The ISO 14001 effort might significantly aid loan issuers in assessing environmental risk prior to commitment and, to a lesser extent, during post-commitment monitoring. As it relates to any plant and/or equipment extension of credit, the emphasis on ISO 14001 compliance information could be integrated into current environmental due diligence processes. One should be aware of the bankers' demand for reliable and comparable data that enables them to analyze similar sorts of financing transactions when determining the precise role ISO 14000 may play. Firms can prove they have a clear plan to eliminate environmental pollution and are integrating environmental management into the entire management of the organization by adhering to the ISO standard. The degree to which bankers believe that this accreditation correlates with genuine financial risk reductions will determine the banking industry's interest in it. The effectiveness of the external information generated by the ISO 14001 process, which is currently being developed, will unquestionably affect this connection. An informal coalition of ten states in the USA conducted an experiment to find out whether EMS are good for the environment and the economy (Campos, 2012).

Setting goals and targets will help you follow the policy statement. Statements like "the establishment of a waste minimization program" should be among the objections. Each purpose and monetary goal would have its own set of targets. The development of the objective and target should begin as soon as feasible and be upgraded consistently, much like the policy statement. To accomplish the goals within a given timeframe, specific and quantifiable environmental targets should be established. Environmental performance indicators are used to measure things like the amount of energy or raw materials used, the amount of a specific compound released into the air, the amount of waste produced in relation to the amount of finished goods produced, the percentage of waste that is recycled, or the number of environmental incidents and accidents. People make decisions in the marketplace every day that have an impact on the environment, either directly or indirectly. Manufacturers can select from a variety of products, suppliers, and production techniques. For their part, consumers decide which products to use or whether they will use a product at all. People who want to make environmentally conscious decisions need accurate information, frequently related to life cycle assessment. A product or service system's entire life cycle, from the obtaining of raw materials through production, usage, and disposal is taken into account by LCA (Life cycle assessment). This knowledge is crucial and can greatly aid in decision-making in business, government, or non-governmental groups by pointing out methods to improve a product's environmental features at different points in its life cycle. The choice of various environmental performance indicators and the promotion of product or service marketing can both benefit greatly from it. The most crucial element of an LCA is probably the defining of the goals. The inventory is a qualitative and/or quantitative analysis of the resources consumed and the emissions produced throughout the course of a product's life cycle. The three components of the effect assessment are classification, characterization, and valuation. Human health, ecological health, and resource utilization are the three basic areas into which the analyzed impacts fall. Characterization involves grouping inventory data into impact categories and applying equivalence criteria to do so. The analysis of the relative contribution of the various inputs or outputs by category is generally quantitative in nature.

Expert panels or the comparison of environmental loading profiles can perform valuation qualitatively or statistically, respectively. Interpretation refers to findings drawn from evaluations and recommendations for corrective measures. International Standards are essential for national production, international terminology, safety and health protection, measurements, analysis, quality control, and environmental protection in all industries. This is especially true in the energy sector, where standards for the points where energy flows meet other forms of energy, such as electrical connectors, fuelling devices, calibration techniques, and electrical safety, are essential. These standards enable integrated resource planning, which incorporates all new renewable options, including co-generation, hybrid systems, small decentralized units, bioenergy, ambient temperature use by heat pumps, and substitutions of powered systems or vice versa, in addition to the more effective production and use of traditional finite and renewable energy sources. Direct and indirect solar energy, for example, is becoming increasingly important. Using this new standard tool in conjunction with the many existing and developing standards on specific energy systems or components thereof, it is possible to calculate the total cost of their emissions, or re-usable embedded energy, as well as their total and relative efficiencies. It is also possible to calculate their life cycles and assess their risks. The majority of the original LCA research evaluated various packaging materials (Huang & Ma, 2004). For instance, numerous studies comparing single-use cartons with disposable and refillable glass and polycarbonate bottles of milk have been conducted. Mechanical recycling is combined with effective feedstock recycling and energy methods in an environmentally friendly recycling and recovery of the packaging material plan.

Chapter

7 CHALLENGES IN WASTE MANAGEMENT

Content

7.1 INTRODUCTION

Losses and waste of agricultural products and food that happen in the agrifood supply chains for various reasons and in varied ways are to be seen as an obvious representation of inefficiencies that characterize all the different production steps of the supply chains themselves. In fact, it is evident that the expenses associated with waste and losses should be considered in addition to the value of the relevant items as well as the environmental and social issues they raise. Although attempts to measure waste have only recently been made, the problem has long been understood. In this regard, FAO's effort in 2011 to quantify the economic and environmental effects of food losses and waste was the first and most significant effort. The United States Department of Agriculture, which views food waste as a part of the larger issue of food loss, has produced another definition of the complicated phenomenon of food losses and waste.

The value of the products that are of interest is significantly influenced by the underlying difference between the problem of food waste in developed and developing nations, which is greater for food waste than for loss of primary products.

The FAO estimates that the losses and waste that happen along the various food supply chains have an annual cost of around 750 billion dollars, based on the prices at which the targeted goods were produced (Ishangulyyev, Kim, & Lee, 2019). Since real values tend to increase in the supply chain and can't be computed using the price at production, this technique of computation, as the FAO itself notes, yields a negative estimate of the real values. Studies have been done specifically on the situations in each country in order to attempt and better quantify waste economically. According to estimates, the cost of products wasted in the UK can range from 1,500 to almost 3,000 dollars per tonne, depending on whether the waste is generated during processing or during final consumption. The USA, where garbage was predicted to cost $2,700 per tonne, came to a similar conclusion.

However, the estimated global cost is higher than that estimated by the FAO by at least 400 billion dollars a year, calculated on the estimated value per tonne of food wasted: $2,500 for the USA, Canada, and Europe; $1,000 for industrialized Asian countries; and $550 dollars for all other areas. This is because retail prices rather than prices at production, as considered by the FAO, of the products interested in losses and waste are taken into account.

7.2 FOOD WASTE

Due to its sheer magnitude, the problem of losses and waste can have an impact on other crucial concerns like food access and environmental sustainability, but it should not be seen as the cause of the criticality of the current agrifood systems; rather, it is one of their effects. In this context, the issue of losses and waste is seen as a component of a highly intricate situation. The way in which the relationships between the various agrifood supply chain components have changed, as well as the obvious inefficiencies and distortions that have resulted, are typical elements. Losses and waste are obvious manifestations of these problems' environmental, social, and production issues. Undoubtedly, the rise in agricultural production has helped feed a growing global population that is becoming more and more urbanized and employs fewer people in agriculture. However, it is undeniable that the agro-industrial models used up to now to assist agricultural development have led to significant changes in agriculture's traditional traits as well as a significant increase in production. Agriculture in particular has been obliged to adapt to industrial logic, streamlining its operations and production methods, which increasingly tend to focus on a single commodity and a single farming method. It is no coincidence that one of the key features of agro-industrial models is that they are based on monoculture, which is itself defined by a high degree of uniformity, both in terms of the factors used and the results obtained. Agriculture has thus been compelled to lose its primary characteristic and requirement, namely to base its production result on a balanced use of existing resources rather than exploitation, as well as to become decontextualized from its environment.

7.3 ECONOMIC ASPECTS

The physiological economic decline of agriculture in sophisticated economic systems has been influenced by a number of variables, the first of which is the tendency toward concentration, which is notably strong in industry and distribution but virtually non-existent in agriculture. Within agrifood supply chains, which can only be compared to pure competition in agricultural production, there are a variety of certainties that coexist side by side. The situation is significantly different for the market for inputs and finished goods, where concentration has frequently been so great as to give rise to oligopolies. Consider that six multinational corporations currently control 80% of the global agrochemicals market and 90% of the seed market, respectively; 15% of the world's largest retail channels control more than

30% of the market for food products; and 20 food industries produce about 20% of packaged food products (Howard, 2015). Since many oligopolistic groups prefer to retain their success, rather than spreading it outside in the form of lower prices for inputs or finished goods, a significant weakening of the spread of development is unavoidable in this context. So, it is expected that there would be distortions, which will lead to inefficiencies and waste. It is also obvious that the price of the goods people purchase includes the cost of waste and losses. It is exceedingly challenging to put together and implement an effective action against waste in a setting like this, where subjects with very varied bargaining power are present within the same agrifood supply chain. For such an activity to be successful, all participants within the various supply chain phases themselves must be involved. Due to the aforementioned, situations where agrifood systems are dominated by significant groups that think it convenient to achieve that goal seem to better adapt to the opportunity of adopting waste reducing programs. For instance, in Europe, the primary initiatives to reduce waste have been either started by businesses that control a significant portion of the food markets in the target countries or by governments of nations with established agrifood systems that have a majority of large groups.

Examples include the initiatives of a multinational company in France engaged in large-scale food service, some organizations engaged in large-scale distribution and public institutions in Great Britain, or a group of food industries in Norway. Governments in the Netherlands and Denmark have also started programs. Launching anti-waste campaigns is more challenging in nations with fragmented production systems since there are many small and medium-sized businesses operating at different stages of the agrifood supply chains, making coordination difficult. As a result, it is significantly more difficult to take action in developing nations where losses rather than waste constitute the primary issue and where altering industrial structures and infrastructure is required to stop this issue. Ironically, the rise of oligopolies, one of the major factors contributing to uneconomical distortions and inefficiencies, becomes one of the finest ways to address the issues it has sparked, at least in more developed nations. This should prompt discussions on the limitations shown by the current agrifood models and, more broadly, by the current development models. It would be appropriate to think specifically about how difficult it is to even postulate alternatives to the current ones and how they may be seen as consistent with every person's right to food sovereignty. The inefficiencies in the use of resources and product distribution are undoubtedly one of the major weaknesses of agricultural production systems. The issue of waste and losses in this scenario is crucial in identifying the system's primary issues, including sustainability, hunger, and the effectiveness of the manufacturing processes. In light of the fact that approximately 40% of agricultural production is either used as animal feed or for purposes other than food, and roughly 30% of food is lost or wasted, it follows that the problem of waste both contributes to and simultaneously is a direct expression of what is, unquestionably, the most serious contradiction of our time (Girotto, Alibardi, & Cossu, 2015).

After taking all of the aforementioned factors into account, one of the first things to remember is that, much like non-food products, food products that are wasted or lost in production require land, water, capital, labor, and, more generally, power. The FAO estimates that around 28% of agricultural land is used for crops whose products are wasted. The vast majority of it is used for farming; if it were a single nation, it would be the second largest in the world. This is already a serious issue, but it becomes even more so when you consider that each year almost 12 million hectares of land are lost due to various forms of depletion worldwide, an area that is roughly equivalent to the amount of farmland used in Italy. Agriculture has tripled its output globally during the past 50 years, despite only a 12 percent increase in farmland. This indicates that soil exploitation has significantly risen, with intensive farming being one of the primary contributors to soil depletion. It should be noted that soil should be observed as a non-renewable resource due to the long time required for its production and taking into account our span. According to estimates, soil loss is greater than 30 times the sustainability rate. Because it is highly likely that the effects of the two phenomena, loss of products and loss of soil, add up, it follows that the use of land for crops whose products are lost should be taken into account not only in terms of the use of soil but also in relation to the type of agriculture practiced and its impact on soil depletion. It is obvious that using land for crops whose products are lost causes further harm by contributing to land loss that could have been prevented, especially in intensive farming or farming that is not sustainability-oriented.

7.4 WATER SAVING

The use of water can be taken into account in a similar manner. According to data from the FAO's water information system, agriculture uses 69 percent of the fresh water used globally; the remaining 19 percent is used by industries, and the remaining 12 percent is used by towns. Due to the prevalence of surface irrigation, such consumption rates are especially high in Africa and Asia, where there is also record waste. 280 million of the 324 million hectares that are irrigated today employ surface irrigation; rice fields make up 29% of the world's irrigated land, while China and India account for 42% of the world's irrigated area. The issue of waste and resources stands heavily on this alarming scenario because, according to the FAO, it is estimated that roughly 9% of the water utilized in agriculture is used for crops whose products are lost or discarded. If we were to attribute the usage of water for crop irrigation in this situation, where the products are wasted, to a single nation, that nation would rank No.1 in the world for irrigation water use. The fact that 61 percent of all irrigated land is used for grain crops, 10 percent for vegetables, and 6 percent for fruit must be emphasized. These sectors suffer losses of 45 percent and 35 percent of their global productions, respectively. It is evident in this instance as well how the loss of agricultural products results in the waste of water that could have been conserved or used in another way. Along with the waste of natural resources, there must also be capital considerations for the production of food and agriculture. The energy required to produce food and agricultural goods, which amounts to roughly 8 gigajoules in developing nations and 35 gigajoules per person annually in Organisation for Economic Co-operation and Development (OECD) nations, can be an interesting component for synthesis. It should be emphasized that in OECD countries, the step of industrial transformation and distribution accounts for half of the aforementioned power usage, while in underdeveloped nations, cooking consumes the other half. Agrifood supply chains as a whole consume roughly 30% of global power, of which 70% is consumed in the processes that come after farming. According to estimates, losses and waste account for around 38 percent of the overall power consumed by agrifood supply chains, or 11.4 percent of the global power supply (Malak-Rawlikowska et al., 2019).

It is important to note that both the use of pesticides and fertilizers has increased when it comes to the chemical components used in the production process, and consequently to production factors that, in the case of lost or wasted productions, represent a useless release of pollutants in the environment. In particular, the pesticide industry increased from a little under 46 billion dollars in 1990 to over 55 billion dollars in 2014, and it is predicted to continue growing until 2018, when it should reach 62 billion dollars. Asia and Latin America regions account for the majority of market shares and over half of global consumption. These regions also experience a considerable loss of agricultural products. Fertilizer usage has grown dramatically. The combined consumption of the macronutrients increased from 161.8 to 190.7 million tonnes between 1990 and 2014 and is expected to continue to rise until 2018 when it is expected to reach 200.5 million tonnes. It must be emphasized that estimates of increased use through 2018 are especially high in the continents that experience the greatest losses in agricultural output, namely Africa and Asia, with peaks in Sub-Saharan regions and South-East Asia. In addition to the aforementioned factors, the impact of losses and waste on the manufacturing processes of agrifood supply chains must also be taken into account with regard to greenhouse gas emissions. It is important to keep in mind that the amount of emissions varies significantly from product to product, depending on where in the supply chain it is generated, as well as the food and production systems of each individual nation. Nevertheless, the emissions attributable to the production of a tonne of vegetable products are significantly lower than those attributable to the production of a similar amount of red meat. In connection to their primary productions and consumption, similar considerations can be made for the individual countries.

In terms of undernourishment, the issue of losses and waste is pertinent for all four of the fundamental elements of food security, namely availability, access, stability, and utilization. It is important to note that malnutrition caused by excess food is concentrated in industrialized nations, where waste is the most important issue, particularly in the retail sale and consuming stages of the agrifood supply chains. Given the growing importance of the issue, excessive food intake and its consequences, such as overweight and obesity, cannot be viewed as the outcome of eating issues. Given that the waste is caused by consumerist culture, which by its very nature can only lead to more waste, even in agrifood supply chains, it is obvious that we are dealing with a complicated situation that demonstrates how pervasive the culture of waste is. It is very challenging to design and implement a single strategy because the problem of losses and waste happens with drastically varying intensity and modality owing to numerous causes, which are linked to the socio-economic characteristics of the interested countries. Instead, plans should be made in light of the numerous certainties they must address. Due to this, the FAO launched "Save Food: Global Initiative on Food Loss and Waste Reduction" in May 2011. This initiative emphasized the necessity of creating and implementing specific programs in collaboration with other international organizations and national governments, involving all parties of the agrifood supply chains, from farmers to multinational corporations. The methodology, which is obviously diverse based on the socio-economic traits of the targeted geographical areas, must be multidisciplinary and integrated due to the sheer number of parties involved. Actions to combat losses and waste are included in more general policies, such as those for sustainable development and the battle against malnutrition, according to the FAO. It should be noted that the elimination of food losses and waste is one of the five goals of the "Zero Hunger Challenge" program, which was launched in 2012 (Lal et al., 2021).

7.5 INTEGRATED SOLUTIONS

A global standpoint and cooperation between public and private parties to implement integrated actions differentiated on a regional basis are unquestionably an important assumption, but they are not absolutely necessary to solve the complex problem of losses and waste. In actuality, this issue must be viewed as primarily a result of the alterations that define agrifood supply networks and of the difficulty in bringing together diverse actors with varying traits and economic weights in order to achieve a single goal. Additionally, this makes government jobs extremely difficult. Governments cannot achieve accomplishments that go beyond fruitless awareness efforts without the active participation of private parties. This condition adds to the complexity of the fight against food losses and waste. In actuality, it is evident that the potential for developing economic interests on which they can converge is inextricably related to the active participation of private parties in initiatives to minimize waste. For developed nations, it is also obvious that such convergence is made simpler with the cooperation between governments and large corporations that operate in the various steps of the supply chains, i.e., parties who, due to their dominant position on the market, are also the most responsible for inefficiencies that later lead to waste, among other things. The main issue in developing nations is food loss, which makes the situation there much more complicated. Finding the optimum point of confluence of public and private interests is particularly challenging in this situation since actions require investments in infrastructures. The investments required to reduce food losses can frequently be compared to investments in development cooperation, a field that has recently been the subject of a contentious discussion about whether private investments in agriculture are consistent with the actual needs and development potential of the countries receiving the investments. It should be noted that many studies have been conducted by international organizations and private and public establishments on the various aspects of the phenomenon of losses and waste due to its complexity and extreme inconsistency, in order to provide useful tools and information to combat the phenomenon, both in developed and developing countries.

We must keep in mind that if the current situation persists, based on anticipated demographic and economic changes, and if the issue of losses and waste is not addressed while taking into account its global significance, it will only get worse, bringing with it the effects that are already manifesting. The predictions made by FAO and OECD beginning in 2009, which have highlighted the need to increase investments in agriculture and food production to meet the needs of a population expected to increase to 9 billion by 2050, are one of the key causes for concern regarding the future of humanity. The demand for raw materials for the making of animal feed and biofuels is also anticipated to increase, as these products are anticipated to be in higher demand as a result of rising incomes in many developing nations and the European Union's decision to have biofuels account for 10% of the global fuel market by 2020. Despite the necessity of planning agricultural development around anticipated demographic growth, it is obvious that such projections fail to adequately account for the fact that sustainability concerns, and thus the responsible use of the resources that are currently available, appear to be a requirement for availability itself rather than concerns regarding the availability of resources. We think that the issue of losses and waste needs to be contextualized in this perspective. We could review the current predictions about the need to increase food production and agricultural output if we significantly reduce what is lost or wasted today. We could also manage the resources we have now more responsibly, reducing the need to intensify or expand our use of them in the coming years.

It should be obvious that since losses and waste are a global issue, only by involving as many parties as possible can they be effectively addressed. This means that efforts to reduce losses and waste can only have a noticeable impact if they can encourage the growth of economic activities and personal behaviors that are more closely related to local resources and, consequently, more aware of their appropriate usage. Building the foundation for economic growth that identifies not only with the results made by large corporations operating in a global market under a regime, but also with the activities that generate widespread welfare through inclusive models based on the enhancement of the resources available, seems to be particularly important and necessary in light of the highlighted criticalities. Only in this way can a new mindset be developed that can help reduce waste and losses with an approach that must be based on knowledge of what is right and in the interest of all, and which, by its very nature, should be the primary requirement to apply and support development models capable of overcoming the limitations and criticality of the current ones.

7.6 ANTIOXIDANTS

Any agent that considerably slows down or stops the oxidation of a substrate when present in quantities below those of the substrate is an antioxidant. Antioxidants are thus an important class of food additives with the capacity to minimize the negative effects of oxygen and nitrogen reactive species in biological systems like the human body. These reactive species can cause harmful changes to food due to the unintentional oxidation of nutrients. Antioxidants function as food additives by impeding oxidative rancidification, preventing the formation of harmful oxidation products, and extending the shelf life of products. One of the main causes of food quality degradation is lipid oxidative rancidification, and adding antioxidants has been found to be the most efficient way to slow it down. In biological systems, antioxidants limit the generation of free radicals, scavenge them, or encourage their breakdown in order to decrease tissue damage caused by free radicals.

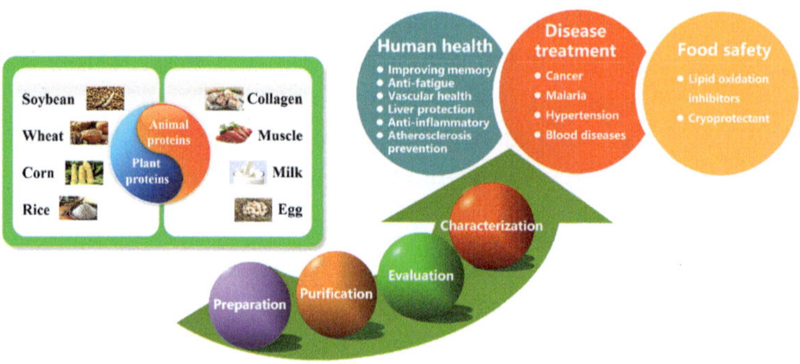

Figure 7.1: Antioxidants in food.

(Source: Pan M, Liu K, Yang J, Liu S, Wang S, Wang S. Advances on Food-Derived Peptidic Antioxidants—A Review. Antioxidants. 2020; 9(9):799. https://doi.org/10.3390/antiox9090799)

There are two types of antioxidants: primary, which are peroxide decomposers like organophosphites and thioesters, and secondary, which are active radical scavengers or chain reaction breakers like polyphenols and secondary aryl amines (Wang, 2000). The prevalent secondary metabolites of plants and found throughout the plant kingdom are polyphenolic compounds, which primarily consist of derivatives and/or isomers of flavones, flavanols, flavanones, isoflavones, anthocyanins, and phenolic acids. They can take the form of simple molecules or high molecular weight polymers. Consumption of these compounds has been linked to a reduced risk of cardiovascular disease or CVD and a decrease in the incidence of human degenerative diseases like cancer. These positive health effects are primarily attributed to their antioxidant activity. Dietary polyphenols are therefore among the most researched compounds during the past decades as components in functional foods and beverages due to the mounting evidence of their potential health benefits. Fruits, vegetables, cereals, coffee, chocolate, wine, and several edible plants are the main sources of these dietary phenolic compounds.

As awareness of the benefits of healthier diets on their quality of life has grown, society's eating habits have changed. As a result of these changes, there is now a higher demand for dietary food products, which has led to a rise in the production of these products over the past few decades. A significant disadvantage of expanding food production, aside from the advantages of appropriately feeding bigger segments of the population, is the parallel increase in the amount of waste produced. The fruit and vegetable industries are estimated to generate wastes in the range of 5 to 50% of the raw material weight, while agricultural practices alone are estimated to generate wastes ranging from 50% to 300 percent of the targeted product weight. The development of food waste has a huge detrimental impact on the environment, society, and the economy. Environmentally speaking, agricultural and food wastes cause volatile organic compound emissions when they are burned for energy recovery and Greenhouse Gas emissions when they are disposed of in landfills. Other environmental effects linked to these wastes include soil, nutrient, water, and energy depletion, as well as the disruption of biogeochemical cycles brought on by intensive farming practices. Economic repercussions are related to the expenses of properly handling wastes and their effects on producer and consumer earnings, whereas social impacts can be attributed to ethical and moral dimensions within the general idea of global food security. The proper disposal of agricultural and food wastes is a serious problem because biological materials often deteriorate quickly due to microbes, reducing their potential for further use. Legal constraints make the problem of waste disposal even worse because, if dumped in soils or water bodies, these pollutants have a high biochemical oxygen requirement for appropriate decomposition and may cause undesirable eutrophication of the surrounding ecosystem. Therefore, it has become increasingly important in recent years to harness these resources in a way that is effective, affordable, and environmentally responsible.

7.7 BIOACTIVE COMPOUNDS

Recovery of bioactive compounds is one of the most promising solutions being researched and put forth for minimizing byproducts in the agricultural and food sectors. These waste types are inexpensive sources of physiologically active compounds, such as polyphenols, and functional food ingredients, such as dietary fibers. Due to existing technologies, these waste types can be recovered and used as functional constituents in food products. The majority of the soft tissues in the processing wastes from fruits and vegetables are rich in antioxidant chemicals and dietary fibers, making them the most extensively researched substrates for the economically advantageous recovery of these components. A sizable body of research has been published in recent years regarding developing methodologies for the quantification and identification of bioactive compounds in agricultural and food wastes, as well as methodologies for their profitable recoveries. At this point, it should be stressed that there is a wealth of literature on the study of recovering polyphenols and other antioxidant compounds from agricultural and food wastes.

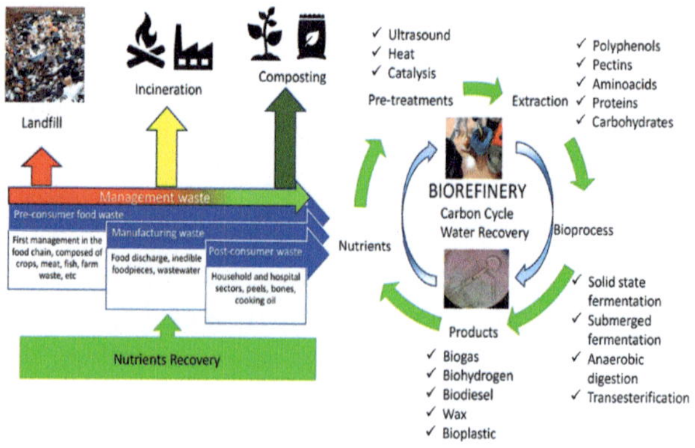

Figure 7.2: Bioactives and waste.

(Source: Ben-Othman S, Jõudu I, Bhat R. Bioactives from Agri-Food Wastes: Present Insights and Future Challenges. Molecules. 2020; 25(3):510. https://doi.org/10.3390/molecules25030510)

Researchers used a variety of bioassays with various molecular targets to examine the antioxidant properties of olive mill wastewater (OMW) high molecular weight phenolic compounds fraction and extracts (Rahmanian, Jafari, & Galanakis, 2014). A clear sample, an ultrafiltrate sample that contained 60% of the total phenolics and was made up primarily of monomeric and oligomeric phenolics, and a sample that contained the remaining 40% of the total phenolic content and was made up entirely of polymeric phenolics were all taken from the OMW sample. Due to the complex phenolic content of the samples, both the ultrafiltrate fraction and the polymeric fraction demonstrated high antioxidant activities in the control of oxidative damage brought on by peroxyl or hydroxyl radicals. Tyrosol and hydroxytyrosol, two monomeric phenolics that makeup around 20% of the total phenolic content of the clear OMW and 30% of the ultrafiltrate fraction, respectively, were found in the clear sample according to the high-performance liquid chromatography (HPLC) analysis results. In the polymeric fraction, neither of these monomeric phenolics was present.

The researchers came to the conclusion that the antioxidant activity of OMW could not be entirely attributed to monomeric phenolics because both ultrafiltrate and polymeric fractions displayed high antioxidant activities. As a result, they chose to recover oligomeric phenolics from clear OMW using membrane filtration technology and to assess their biological activity. Approximately 52% of the total phenolic content of clear OMW was present in the recovered fraction, which also contained tyrosol and hydroxytyrosol. HPLC-MS analysis of this ultrafiltrate fraction also confirmed the presence of verbascoside and isoverbascoside.

To produce an OMW extract free of tyrosol and hydroxytyrosol, the ultrafiltrate fraction was subsequently subjected to chromatographic separation on Sephadex LH-20. The HPLC-MS/MS analysis of this fraction showed the presence of phenolic compounds with molecular weights between 600 and 1000 Da (Cardinali et al., 2010).

In addition to verbascoside and isoverbascoside, the LH-20 fraction analysis also found the existence of verbascoside that has undergone oxidation, as well as a number of higher molecular weight phenolics formed by the oxidative polymerization of hydroxytyrosol and caffeic acid. OMW extracts and regular hydroxytyrosol were measured on human colon cancer cells by utilizing a dichlorodihydrofluorescein diacetate (DCFDA) probe in order to assess the potential protective effects of cells against induced oxidative stress using flow cytometry. Both the ultrafiltrate and the LH-20 fractions showed concentration-dependent antioxidant activity. The ability of high molecular weight phenolics to reduce oxidative stress was also noted to be stronger than that of low molecular weight phenolics.

Hydroxytyrosol was the main phenolic compound in the HPLC-analyzed OMW, while phenolic acids and flavonoids were also found. The antioxidant activity of the extracts formed by solvent and supercritical fluid extractions was evaluated, with the ethanol extract showing the highest antiradical activity. The antiradical activity and phenolic content did not correlate. Despite having a low phenolic concentration, the SFE extract had a decent antioxidant capacity. The Rancimat method on sunflower oil determined that the ethanol extract is a more potent antioxidant than BHT (butylated hydroxytoluene), ascorbyl palmitate, and vitamin E. The phenolic chemicals hydroxytyrosol, tyrosol, and the dialdehydic form of decarboxymethyl elenolic acid, associated with ethanol, made up a phytocomplex, identified from olive mill waste fluids. This phytocomplex makes up 6% of the overall phenolic content. The phytocomplex's antioxidant capacity was assessed using the red blood cell technique, which measures cellular antioxidant activity. It was found that the phytocomplex was able to penetrate the cell membrane and exhibit antioxidant activity there. Using a membrane method and chromatographic purification on Sephadex LH-20, researchers isolated a polyphenolic fraction of OMW, and verbascoside was shown to be the most prevalent phenolic constituent in the LH-20 fraction. The effects of verbascoside against oxidative stress in cultures of human colon cancer cells as well as its capacity to block LDL (low-density lipoprotein) oxidative modification caused by metal catalysts were examined. This fraction was further refined to obtain a verbascoside with a high degree of purity. Verbascoside greatly reduced the cytotoxicity of reactive oxygen species created from H_2O_2 before striking biomolecules in the cell system under study. Verbascoside was integrated in a concentration and time-dependent manner, with an accumulation efficiency of roughly 0.12%. Its strong affinity for negatively charged phospholipid membranes was partially responsible for its antioxidant actions in cellular systems. It was shown that verbascoside may shield LDL from copper-catalyzed oxidation. This ability was attributed to the ability of verbascoside as an electron donor rather than to its capacity to chelate divalent copper. The antioxidant capacity of verbascoside was 2 to 10 times greater than that of the LH-20 fraction as a whole, which includes tyrosol and hydroxityrosol, when all antioxidant assays used in the study were taken into account. As a result, verbascoside was recognized as a powerful biophenol antioxidant with interesting uses in the nutraceutical, cosmetic, and functional food industries.

Tyrosol, caffeic acid, gallic acid, p-coumaric acid, oleuropein, and its aglycone were additional phenolics (McDonald, Prenzler, Antolovich, & Robards, 2001). While the ethyl acetate extract and the microfiltrate fraction both displayed significant antioxidant capacity, the microfiltrate fraction had greater activity. Since the ethyl acetate extraction is more selective for low and medium molecular weight phenols, the higher activity for the microfiltrate fraction was attributed to its diversified phenolic composition, which includes heavier phenolics that were absent in the ethyl acetate extracts. In order to recover polyphenols from OMW, some assessed the effectiveness of dry *Azolla caroliniana*, a freshwater aquatic fern, and granulated activated carbon. Desorption was accomplished using ethanol at pH 11 and NaOH for GAC and acidified water at pH 3 and HCl for the Azolla adsorbent. This difference was explained by the ability of the pectin found in Azolla to electrostatically interact with the positively charged polyphenol molecules at the pH of the adsorption solution. The polyphenol content of the powder obtained by desorption from the Azolla adsorbent was 2.13 times higher than that obtained by desorption from GAC. The gallic, vanillic, caffeic, syringic, and coumaric acids, as well as hydroxytyrosol and tyrosol, were among the phenolic substances desorbed from Azolla. Gallic acid and tyrosol made up the majority of the Azolla desorption product. The GAC desorption product contained only five biophenols: hydroxytyrosol, tyrosol, gallic acid, syringic acid, and coumaric acid. The GAC desorption powder had more hydroxytyrosol than the Azolla desorption product did, proving GAC has a strong affinity for this tiny molecule. Nevertheless, despite the changes in the polyphenol composition of the desorption products, the antioxidant activity of these products was relatively similar.

The desorption products contain polyphenols, the most powerful of which is hydroxytyrosol, which was identified in higher amounts in the GAC product than the Azolla product. It was discovered that there was a nearly linear association between the polyphenol content and the antiradical activity of both Azolla and GAC desorption powders. For the adsorptive recovery of OMW phenolic compounds, both Azolla and GAC were found to be appropriate matrices. Researchers investigated the phenolic, tocopherol, flavonoid, flavanol, dietary fiber, and beta-carotene contents as well as the antioxidant activity of the olive-waste cakes from the processing of the cultivars Picual, Frantoio, and Arbequina. The free extracts included the polyphenols tyrosol and hydroxytyrosol. The Arbequina variety had the highest concentration of carotene, whereas the Picual variety had the highest free phenolic and free flavonoid contents and, thus, the highest antioxidant potential. Due to an expansion in the usage of red ginseng products in the functional food business, processing companies produce and discard enormous volumes of red ginseng marc as waste. The non-fermented RGM (red grape marc) extracts did not contain any polyphenols, but the polyphenol content of RGM extracts dramatically increased after fermentation. The polyphenol contents of BSRGM and SCRGM were 7.226 and 6.875 g/100 g, respectively, and the fermented extracts had significantly higher radical scavenging abilities than NRGM extracts, indicating a positive relationship between polyphenol content and antioxidant activity, as determined by free radical scavenging abilities. The scientists concluded that the radical scavenging properties of the non-fermented extracts, despite the absence of polyphenols, were caused by the presence of other bioactive substances including uronic and sialic acids.

7.8 MORE ABOUT ISOLATED POLYPHENOLS

The non-edible portion of the onion bulb, which includes the outer dry and semi-dry layers and the apical trimmings, is known as onion solid wastes, and it is estimated that 6,000,000 tonnes of OSW are produced each year in Europe (Voll, 1993). Previous studies using water/ethanol mixtures as solvents showed that OSW extracts were substantially richer in polyphenols than the extracts of the edible analog.

They also showed that the polyphenol composition of OSW was distinct from the polyphenol composition of the edible section of the onion. Despite being non-toxic and having the potential to extract polyphenols, ethanol has a high cost and is disposed of in accordance with state legislation, making its recovery and reuse imperative. In order to recover antioxidant polyphenols from onion solid wastes, researchers presented a green ultrasound-assisted extraction method using an aqueous solution of glycerol as a cheap and non-toxic solvent.

The yield of polyphenols at saturation, at 80°C, with the ultrasound-assisted extraction was 1.14 times higher than that with the pH-regulated extraction of OSW with 60 percent ethanol. When the extraction temperature was raised from 50 to 80°C, it was found that the reducing power, a metric of the antioxidant activity, increased by around 1.6 times. Thus, it was concluded that there was no loss of antioxidants as the extraction temperature was raised and that the higher yield of total polyphenols reached at 80°C might be directly related to the increased antioxidant activity.

Caffeine and other components were first eliminated step by step by elution with a 20 percent ethanol solution, and then tea polyphenols were eliminated step by step by reversed flow elution with a 60 percent ethanol solution(Brunetto et al., 2007). With a purity of around 98 percent and a yield of more than 11 percent, catechins made up more than 91 percent of the total amount of polyphenols. High-performance liquid chromatography proved that caffeine was not found in the prepared extract. About 86 percent of the tea polyphenols were recovered using column chromatography, and the preparation method had excellent stability. The solids left after the fermentation of red grapes and wine are known as wine pomace (WP). WP is mostly made up of solid grape skin, remaining pulp, seeds, and any yeasts or bacteria that were present throughout the fermentation process. WP has the potential to be a source of functional food ingredients due to its abundance of bioactive phytochemicals. Researchers assessed the capacity of wine pomace to prevent oxidative deterioration of foods, hence extending their shelf life, and restricting the proliferation of spoiler microorganisms in light of this. The photodimerization and isomerization of phenolic compounds induced by UV radiation, as well as the potential formation of new linkages between hydroxycinnamic acids and lignin units that would decrease their extractability, were the reason for the reductions in total polyphenol content brought on by the UV treatments. The total polyphenol, catechin, and proanthocyanidin levels of the by-products are strongly associated with results, indicating that their phenolic composition is primarily responsible for their antioxidant activity. The results were thought to correlate well with their estimated antioxidant capabilities from ABTS and, consequently, with their polyphenolic levels in terms of their ability to block fat oxidation. Skins wine pomace product (SkWPP) exerted protection against fat oxidation that was significantly higher than that of SdWPP, despite having a lower total polyphenol content than the seeds wine pomace 79 product (SdWPP) (García-Lomillo et al , 2014). This protection was attributed to the higher content of anthocyanins of SkWPP, which has already been shown to have an excellent ability to scavenge the free radicals formed during fat oxidation.

The primary antioxidant found in oils is tocopherol, and the amount found in each oil varied from 29.2 mg/kg in soursop to 328.7 mg/kg in tomato seed oils. Except for passion fruit seed oil, where the predominant isomers were +/- tocopherol, and soursop seed oil, which had a larger concentration of tocopherol, the important isomer was tocopherol. Catechin in grape seed oil, p-coumaric acid in guava seed oil, salicylic acid in passion fruit seed oil, and epicatechin in soursop seed oil were the predominant phenolic components identified in the oil samples. All oil samples included carotenoids, a class of pigments with the ability to inhibit lipid oxidation. The total carotenoid concentration ranged from 4.4 g/g for soursop seed oil to 11.9 g/g for grape seed oil. For the DPPH technique (1,1-diphenyl-2-picrylhydrazyl), soursop seed oil showed the strongest radical scavenging activity.

Mango industrial processing results in large byproduct production, with 35 to 60 percent of the fruit being wasted and primarily being made up of seeds, peels, and paste. These by-products have the potential to be used as components in functional foods because they are a significant source of dietary fiber and bioactive substances like polyphenols. In order to substitute sugarcane with dried "Ataulfo '' mango by-products as a source of dietary fiber and polyphenols, people produced starch-molded mango snacks. The starch-molded mango snacks that resulted from adding the peel and paste of the dried "Ataulfo" mango byproduct to the mango concentrate were evaluated for their physical and sensory qualities, dietary fiber and polyphenol content, and antioxidant potential. When sugarcane was substituted with 25:75 DAMC: DAMB, the polyphenol content increased by roughly five times compared to the control snack formulation (Blancas-Benitez et al , 2015). Compared to the control snack, the mango by-product-enriched food had increased antioxidant activity. By reducing the sugar level and increasing the dietary fiber and polyphenol content of starch-molded mango snacks, it has been shown that adding dried mango byproducts can improve their polyphenol and dietary fiber contents without significantly changing their physical-chemical features. However, it was determined that sensory properties needed to be improved. Aqueous solutions of ethanol, butanediol, and glycerol were the examined solvents. The extract produced using 50% ethanol showed more reducing power, or better antioxidant activity, than the other extracts, however, all of the investigated solvents functioned well.

In order to compare the functional and antioxidative capabilities of date seed proteins to carnosine, an antioxidative peptide, date seed proteins were hydrolyzed. When hydrolysates were added to fish model systems, oxidation was inhibited more effectively than with the synthetic antioxidant butylated hydroxytoluene. The hydrolysates prevented DNA cleavage brought on by hydroxyl and peroxyl radicals as well as the oxidation of LDL cholesterol and beta-carotene. Date seed protein hydrolysates were identified as a possible functional food ingredient for health promotion in light of the findings. Researchers identified the physicochemical, microbiological, and bioactive characteristics of flours produced from the skin and bagasse of kiwi fruit at two developmental stages. For all of the treatments, there were noticeable changes in the phenolic chemicals found in the kiwi flours, with the skin-derived flours having higher concentrations than the bagasse-derived flour. For both kinds, kiwi fruit skin flour contained more bioactive components and had stronger antioxidant activity than kiwi fruit bagasse flour.

7.9 SAFFRON AND OTHER EXTRACT PROCESSING

Saffron is a spice from the stigmas of *Crocus sativus L.* flowers, and throughout the process of making it, only the stigmas are harvested and used, with the remaining flower parts being thrown away as waste. However, it has been found that the flower's waste materials have a lot of potential for application in the food and health industries since they include bioactive substances like carotenoids, polyphenols, anthocyanins, and flavonoids. These bioactive substances are well-known for having strong antioxidant properties. As a result, researchers used high-resolution mass spectrometry to look into the composition of the polar fraction of the extracts and examined the *in vitro* antioxidant activity of juices made from cold-pressing saffron floral by-products. To assess the impact of post-harvest time on the quality of the floral by-products, total phenol content and antioxidant activity were measured in cold-pressed juices collected 24 and 48 hours after saffron flower harvesting. Juices obtained by cold-pressing the by-products at 48 h after harvesting exhibited an increase of 61, 56, and 76 percent for TP, FRAP and DPPH, respectively, in relation to the values determined for the juices obtained at 24 h after harvesting. These values are around two to three times greater than those for red wine or berry extracts.

The removal of polyphenols by a polyvinylpolypyrrolidone resin, used to stabilize beer and increase its shelf life, results in significant levels of polyphenols in brewery waste streams (Barbosa-Pereira, Aurrekoetxea, Angulo, Paseiro-Losada, & Cruz, 2014). These polyphenols can be recovered from the PVPP by an alkaline treatment. The high amounts of phenolic components, such as catechins, gallic, caffeic, p-coumaric, and ferulic acids, which function as free radical acceptors and chain breakers, are responsible for the alkaline extract's high antioxidant activity. These characteristics led to comparison of the antioxidant effects of PVPP-WS, rosemary extracts, and two synthetic antioxidants on the oxidative stability of beef during storage. The natural extracts were used to cover antioxidant-rich active films. The natural extract made from brewery waste stream showed antioxidant activity that was six times more than that of rosemary extract and seven times greater than that of BHT. By directly applying the various antioxidants to the surface of the beef and incorporating them into a packaging film, the antioxidant effects of the various antioxidants were assessed in samples of beef. Both the PVPP-WS and the rosemary extracts demonstrated their efficacy as antioxidants by significantly reducing lipid oxidation during beef storage when applied directly to the surface of the beef samples. As a result, it was decided that the antioxidant extracts must come into direct contact with food. The beef's oxidative stability was improved by the produced active films coated with natural extracts when compared to that obtained with the film control during cold storage. The extract from the brewery residual stream produced the best results, with lipid oxidation being reduced by up to 90%.

The industry that processes crustaceans generates large amounts of waste each year, most of which are made up of cephalothorax and cuticles from peeling, damaged muscle, and cooking liquids. One option for successful use of this waste is the recovery of biopolymers to create edible films or coatings for food packaging purposes. Scientists developed and characterized shrimp muscle protein antioxidant films incorporating carotenoids from a shrimp-waste extract rich in astaxanthin and from a tomato peel waste extract rich in lycopene, taking into account that cephalothorax and cuticles from shrimp waste contain other bioactive compounds of interest, such as astaxanthin, and that the tomato industry generates significant amounts of waste rich in carotene. Between the lipid extracts from tomato peels and shrimp muscle proteins, as well as commercial beta-carotene, there were notable changes in antioxidant activity, with the lipid extracts from food waste exhibiting higher activities. The inclusion of additional antioxidants in the lipid extracts, such as tocopherol and hydrophobic amino acids, was thought to be the cause of these variations The tomato-extract, commercial carotene extract, and shrimp-extract films exerted the next-highest antioxidant activity, respectively. Astaxanthin demonstrated the slowest storage-related deterioration among the investigated carotenoids, which was utilized as a metric to examine the stability of the generated films. As storage continued, a progressive decline in the amount of released carotenoids was seen. This decline was consistent with the rate of degradation seen for each carotenoid, with carotene experiencing a faster and more significant decline than astaxanthin. The results for carotenoid release from the tomato-extract film, however, did not match those for carotenoid degradation; the discrepancy was attributed to a potential interaction between lycopene and the film matrix, which might have prevented carotenoid release. However, lycopene degradation was also noted.

The existence of mono and diesters rather than free astaxanthin and the higher antioxidant activity of the shrimp extract were both thought to contribute to its higher stability (Gómez-Estaca, Calvo, Álvarez-Acero, Montero, & Gómez-Guillén, 2017). Since commercial carotene and tomato-extract lycopene were both present in the free form, this made them more likely to degrade and interact with the proteins in the film. The antioxidant activity of the films was unaffected by the carotenoid molecule degradation, and it remained steady during the course of storage. The presence of additional antioxidant compounds in both shrimp and tomato extracts as well as the presence of degradation products with antioxidant activity, which was inferred from the stable behavior of the commercial carotene film, were then attributed to the stability of the antioxidant activity during its storage. As a result, it was decided that the formation of edible antioxidant films containing lipid extracts rich in carotenoids was a practical way to make use of shrimp and tomato waste products. A barley bran protein-gelatin composite film that contains an antioxidant and antibacterial grapefruit seed extract was created and effectively used for salmon packaging. Researchers evaluated and successfully confirmed the efficacy of an active packaging film containing antioxidants extracted from barley husks, an agricultural waste, to protect salmon, blue shark, and cod against oxidative lipid degradation during storage in a number of studies. Astaxanthin, a by-product of shrimp production, was successfully incorporated into a matrix of low-density polyethylene.

The reduction of MDA (malondialdehyde) and hexanal by spent coffee grounds, whole grounds, and lyophilized brew in frozen cooked pork patties was comparable to that of rosemary oleoresin, showing that coffee has a lot of potential as a substitute antioxidant for preventing lipid oxidation. No matter how it was added, coffee had no effect on the pork protein oxidation as indicated by the production of metmyoglobin or free thiol content. The shelf life of the frozen product could be increased by adding either type of coffee roast to the cooked pork patties. Due to their biodegradability and accessibility, organic compounds recovered from food waste have thus been considered as substitute inhibitors.

7.10 MORE RESEARCH ON EXTRACTS

Several food waste extracts, including fruit and garlic peel extracts, have been successfully tested as corrosion inhibitors of steel in acidic media. The corrosion inhibition ability of plant extracts is typically attributed to the presence of antioxidant polyphenolic compound constituents like alkaloids, flavonoids, or condensed tannins. Orange peel, a byproduct of the juice extraction process, makes up about 50% of the total fruit mass and is a major source of phenolic compounds with substantial antioxidant activity. The HPLC analysis of the orange peel extracts revealed the presence of ascorbic acid, neohesperidin, hesperidin, nobiletin, naringin, and narirutin as the major phenolic components. The orange peel extract and the major individual compounds found in the extract were both tested for antioxidant activity(M'hiri et al., 2016). In contrast to what would be predicted by adding the individual contributions of the main phenolic components, the orange peel extract demonstrated stronger antioxidant activity. The existence of naturally occurring antioxidants, such as carotenoids, and synergistic interactions between various antioxidant molecules, particularly flavonoids and ascorbic acid, were thought to be responsible for this variation in antioxidant activity. Electrochemical techniques were used to examine the effectiveness of the orange peel extract, particular phenolic components, and ascorbic acid in preventing corrosion. This effectiveness was compared to the antioxidant activity measured by the ABTS assay. Orange peel extract showed a strong corrosion inhibition regardless of dose, but some antioxidant components showed just a moderate inhibitory effect. Along with observations made using scanning electron microscopy following immersion, the electrochemical spectroscopy data demonstrated that the orange peel extract's ability to inhibit was also a result of the precipitation of a surface coating. The antioxidant activity displayed by the presence of ene-diol or catechol functions or other derived functions was partially responsible for the cathodic reaction's inhibition. Due to their chelating abilities toward metallic cations, anodic inhibition was linked to the same kinds of functions. By gradually precipitating a covering film made of substances derived from the orange peel, which could cover the steel surface and entrap the antioxidant molecule chelating the Fe^{2+} cations, the corrosion rate was decreased in the presence of the extract. Used coffee grounds are roasted and ground coffee that has lost most of its water solubles.

Hence, to summarize, an improvement in the level of green-technology innovation is warranted. This necessitates the relevant policy

Additionally, the industry and academics should boost technological cooperation and exchanges in such innovations. The research and development of clean technologies for this planet also need to be increased.

REFERENCES

1. Abdulmalek, S. A., Li, K., Wang, J., Ghide, M. K., & Yan, Y. (2021). Co-Immobilization of Rhizopus oryzae and Candida rugosa Lipases onto mMWCNTs@4-arm-PEG-NH2—A Novel Magnetic Nanotube–Polyethylene Glycol Amine Composite—And Its Applications for Biodiesel Production. *International Journal of Molecular Sciences*, *22*(21), 11956. https://doi.org/10.3390/ijms222111956
2. Ahut Daggash, H., & Dowell, N. M. (2021). Delivering low-carbon electricity systems in sub-Saharan Africa: Insights from Nigeria. *Energy & Environmental Science*, *14*(7), 4018–4037. https://doi.org/10.1039/D1EE00746G
3. Ananno, A. A., Masud, M. H., Chowdhury, S. A., Dabnichki, P., Ahmed, N., & Arefin, A. Md. E. (2021). Sustainable food waste management model for Bangladesh. *Sustainable Production and Consumption*, *27*, 35–51. https://doi.org/10.1016/j.spc.2020.10.022
4. Angus, A., Burgess, P. J., Morris, J., & Lingard, J. (2009). Agriculture and land use: Demand for and supply of agricultural commodities, characteristics of the farming and food industries, and implications for land use in the UK. *Land Use Policy*, *26*, S230–S242. https://doi.org/10.1016/j.landusepol.2009.09.020
5. Aravani, V. P., Sun, H., Yang, Z., Liu, G., Wang, W., Anagnostopoulos, G., … Papadakis, V. G. (2022). Agricultural and livestock sector's

residues in Greece & China: Comparative qualitative and quantitative characterization for assessing their potential for biogas production. *Renewable and Sustainable Energy Reviews*, *154*, 111821. https://doi.org/10.1016/j.rser.2021.111821

6. Aulakh, M. S., & Malhi, S. S. (2005). Interactions of Nitrogen with Other Nutrients and Water: Effect on Crop Yield and Quality, Nutrient Use Efficiency, Carbon Sequestration, and Environmental Pollution. In *Advances in Agronomy* (Vol. 86, pp. 341–409). Academic Press. https://doi.org/10.1016/S0065-2113(05)86007-9
7. Ayodele, T. R., Ogunjuyigbe, A. S. O., & Alao, M. A. (2017). Life cycle assessment of waste-to-energy (WtE) technologies for electricity generation using municipal solid waste in Nigeria. *Applied Energy*, *201*, 200–218. https://doi.org/10.1016/j.apenergy.2017.05.097
8. Baiano, A. (2014). Recovery of Biomolecules from Food Wastes—A Review. *Molecules*, *19*(9), 14821–14842. https://doi.org/10.3390/molecules190914821
9. Bajwa, D. S., Peterson, T., Sharma, N., Shojaeiarani, J., & Bajwa, S. G. (2018). A review of densified solid biomass for energy production. *Renewable and Sustainable Energy Reviews*, *96*, 296–305. https://doi.org/10.1016/j.rser.2018.07.040
10. Barbieri, P., Pellerin, S., Seufert, V., & Nesme, T. (2019). Changes in crop rotations would impact food production in an organically farmed world. *Nature Sustainability*, *2*(5), 378–385. https://doi.org/10.1038/s41893-019-0259-5
11. Barbosa-Pereira, L., Aurrekoetxea, G. P., Angulo, I., Paseiro-Losada, P., & Cruz, J. M. (2014). Development of new active packaging films coated with natural phenolic compounds to improve the oxidative stability of beef. *Meat Science*, *97*(2), 249–254. https://doi.org/10.1016/j.meatsci.2014.02.006
12. Bhawangirkar, D. R., & Sangwai, J. S. (2019). Insights into Cage Occupancies during Gas Exchange in CH4+CO2 and CH4+N2+CO2 Mixed Hydrate Systems Relevant for Methane Gas Recovery and Carbon Dioxide Sequestration in Hydrate Reservoirs: A Thermodynamic Approach. *Industrial & Engineering Chemistry Research*, *58*(31), 14462–14475. https://doi.org/10.1021/acs.iecr.9b02364
13. Biswas, A. K., Kumar, S., Babu, S. S., Bhattacharyya, J. K., & Chakrabarti, T. (2010). Studies on environmental quality in and around municipal solid waste dumpsite. *Resources, Conservation and Recycling*, *55*(2), 129–134. https://doi.org/10.1016/j.resconrec.2010.08.003

14. Blancas-Benitez, F. J., de Jesús Avena-Bustillos, R., Montalvo-González, E., Sáyago-Ayerdi, S. G., & H. McHugh, T. (2015). Addition of dried 'Ataulfo' mango (*Mangifera indica* L) by-products as a source of dietary fiber and polyphenols in starch molded mango snacks. *Journal of Food Science and Technology*, *52*(11), 7393–7400. https://doi.org/10.1007/s13197-015-1855-7
15. Boudouropoulos, I. D., & Arvanitoyannis, I. S. (2000). Potential and Perspectives for Application of Environmental Management System (ems) and Iso 14000 to Food Industries. *Food Reviews International*, *16*(2), 177–237. https://doi.org/10.1081/FRI-100100287
16. Brix, H. (1993). Wastewater Treatment in Constructed Wetlands: System Design, Removal Processes, and Treatment Performance. In *Constructed Wetlands for Water Quality Improvement*. CRC Press.
17. Brunetto, M. del R., Gutiérrez, L., Delgado, Y., Gallignani, M., Zambrano, A., Gómez, Á., … Romero, C. (2007). Determination of theobromine, theophylline and caffeine in cocoa samples by a high-performance liquid chromatographic method with on-line sample cleanup in a switching-column system. *Food Chemistry*, *100*(2), 459–467. https://doi.org/10.1016/j.foodchem.2005.10.007
18. Campos, L. M. S. (2012). Environmental management systems (EMS) for small companies: A study in Southern Brazil. *Journal of Cleaner Production*, *32*, 141–148. https://doi.org/10.1016/j.jclepro.2012.03.029
19. Cao, G., Zhang, X., & Zheng, F. (2006). Inventory of black carbon and organic carbon emissions from China. *Atmospheric Environment*, *40*(34), 6516–6527. https://doi.org/10.1016/j.atmosenv.2006.05.070
20. Cardinali, A., Cicco, N., Linsalata, V., Minervini, F., Pati, S., Pieralice, M., … Lattanzio, V. (2010). Biological Activity of High Molecular Weight Phenolics from Olive Mill Wastewater. *Journal of Agricultural and Food Chemistry*, *58*(15), 8585–8590. https://doi.org/10.1021/jf101437c
21. Chen, M.-F. (2008). Consumer Trust in Food Safety—A Multidisciplinary Approach and Empirical Evidence from Taiwan. *Risk Analysis*, *28*(6), 1553–1569. https://doi.org/10.1111/j.1539-6924.2008.01115.x
22. Chen, W.-H., Lu, K.-M., & Tsai, C.-M. (2012). An experimental analysis on property and structure variations of agricultural wastes undergoing torrefaction. *Applied Energy*, *100*, 318–325. https://doi.org/10.1016/j.apenergy.2012.05.056

23. Chhabra, A., Manjunath, K. R., Panigrahy, S., & Parihar, J. S. (2013). Greenhouse gas emissions from Indian livestock. *Climatic Change*, *117*(1), 329–344. https://doi.org/10.1007/s10584-012-0556-8
24. Cieciura-Włoch, W., Borowski, S., & Otlewska, A. (2020). Biohydrogen production from fruit and vegetable waste, sugar beet pulp and corn silage via dark fermentation. *Renewable Energy*, *153*, 1226–1237. https://doi.org/10.1016/j.renene.2020.02.085
25. Clark, G. (2007). Evolution of the global sustainable consumption and production policy and the United Nations Environment Programme's (UNEP) supporting activities. *Journal of Cleaner Production*, *15*(6), 492–498. https://doi.org/10.1016/j.jclepro.2006.05.017
26. Correia, M. I. T. D., Hegazi, R. A., Higashiguchi, T., Michel, J.-P., Reddy, B. R., Tappenden, K. A., … Muscaritoli, M. (2014). Evidence-Based Recommendations for Addressing Malnutrition in Health Care: An Updated Strategy From the feedM.E. Global Study Group. *Journal of the American Medical Directors Association*, *15*(8), 544–550. https://doi.org/10.1016/j.jamda.2014.05.011
27. Coudard, A., Corbin, E., de Koning, J., Tukker, A., & Mogollón, J. M. (2021). Global water and energy losses from consumer avoidable food waste. *Journal of Cleaner Production*, *326*, 129342. https://doi.org/10.1016/j.jclepro.2021.129342
28. Dai, A. (2011). Drought under global warming: A review. *WIREs Climate Change*, *2*(1), 45–65. https://doi.org/10.1002/wcc.81
29. de Lange, W., & Nahman, A. (2015). Costs of food waste in South Africa: Incorporating inedible food waste. *Waste Management*, *40*, 167–172. https://doi.org/10.1016/j.wasman.2015.03.001
30. de Nardi, I. R., Del Nery, V., Amorim, A. K. B., dos Santos, N. G., & Chimenes, F. (2011). Performances of SBR, chemical–DAF and UV disinfection for poultry slaughterhouse wastewater reclamation. *Desalination*, *269*(1), 184–189. https://doi.org/10.1016/j.desal.2010.10.060
31. de Souza, R. M., Seibert, D., Quesada, H. B., de Jesus Bassetti, F., Fagundes-Klen, M. R., & Bergamasco, R. (2020). Occurrence, impacts and general aspects of pesticides in surface water: A review. *Process Safety and Environmental Protection*, *135*, 22–37. https://doi.org/10.1016/j.psep.2019.12.035
32. Demirbas, A. (2008). Biomethanol Production from Organic Waste Materials. *Energy Sources, Part A: Recovery, Utilization,*

and Environmental Effects, *30*(6), 565–572. https://doi.org/10.1080/15567030600817167

33. Demirbas, Ayhan. (2011). Waste management, waste resource facilities and waste conversion processes. *Energy Conversion and Management*, *52*(2), 1280–1287. https://doi.org/10.1016/j.enconman.2010.09.025
34. Demirel, B., & Yenigun, O. (2004). Anaerobic acidogenesis of dairy wastewater: The effects of variations in hydraulic retention time with no pH control. *Journal of Chemical Technology & Biotechnology*, *79*(7), 755–760. https://doi.org/10.1002/jctb.1052
35. Derbal, K., Bencheikh-lehocine, M., Cecchi, F., Meniai, A.-H., & Pavan, P. (2009). Application of the IWA ADM1 model to simulate anaerobic co-digestion of organic waste with waste activated sludge in mesophilic condition. *Bioresource Technology*, *100*(4), 1539–1543. https://doi.org/10.1016/j.biortech.2008.07.064
36. Doelle, M., & Sinclair, A. J. (2006). Time for a new approach to public participation in EA: Promoting cooperation and consensus for sustainability. *Environmental Impact Assessment Review*, *26*(2), 185–205. https://doi.org/10.1016/j.eiar.2005.07.013
37. Domínguez-Bocanegra, A. R., Torres-Muñoz, J. A., & López, R. A. (2015). Production of Bioethanol from agro-industrial wastes. *Fuel*, *149*, 85–89. https://doi.org/10.1016/j.fuel.2014.09.062
38. Dutta, T., Kwon, E., Bhattacharya, S. S., Jeon, B. H., Deep, A., Uchimiya, M., & Kim, K.-H. (2017). Polycyclic aromatic hydrocarbons and volatile organic compounds in biochar and biochar-amended soil: A review. *GCB Bioenergy*, *9*(6), 990–1004. https://doi.org/10.1111/gcbb.12363
39. Emamjomeh, M. M., Tahergorabi, M., Farzadkia, M., & Bazrafshan, E. (2018). A Review of the Use of Earthworms and Aquatic Worms for Reducing Sludge Produced: An Innovative Ecotechnology. *Waste and Biomass Valorization*, *9*(9), 1543–1557. https://doi.org/10.1007/s12649-017-9907-z
40. Esposito, G., Frunzo, L., Giordano, A., Liotta, F., Panico, A., & Pirozzi, F. (2012). Anaerobic co-digestion of organic wastes. *Reviews in Environmental Science and Bio/Technology*, *11*(4), 325–341. https://doi.org/10.1007/s11157-012-9277-8
41. Evans, E. A., Ballen, F. H., & Siddiq, M. (2017). Mango Production, Global Trade, Consumption Trends, and Postharvest Processing and

Nutrition. In *Handbook of Mango Fruit* (pp. 1–16). John Wiley & Sons, Ltd. https://doi.org/10.1002/9781119014362.ch1

42. Ferraz de Campos, V. A., Silva, V. B., Cardoso, J. S., Brito, P. S., Tuna, C. E., & Silveira, J. L. (2021). A review of waste management in Brazil and Portugal: Waste-to-energy as pathway for sustainable development. *Renewable Energy, 178*, 802–820. https://doi.org/10.1016/j.renene.2021.06.107
43. Ferronato, N., Portugal Alarcón, G. P., Guisbert Lizarazu, E. G., & Torretta, V. (2021). Assessment of municipal solid waste collection in Bolivia: Perspectives for avoiding uncontrolled disposal and boosting waste recycling options. *Resources, Conservation and Recycling, 167*, 105234. https://doi.org/10.1016/j.resconrec.2020.105234
44. Fischer, G., Prieler, S., van Velthuizen, H., Berndes, G., Faaij, A., Londo, M., & de Wit, M. (2010). Biofuel production potentials in Europe: Sustainable use of cultivated land and pastures, Part II: Land use scenarios. *Biomass and Bioenergy, 34*(2), 173–187. https://doi.org/10.1016/j.biombioe.2009.07.009
45. Frewer, L. J., & Gremmen, B. (2007). 2—Consumer interests in food processing waste management and co-product recovery. In K. Waldron (Ed.), *Handbook of Waste Management and Co-Product Recovery in Food Processing* (pp. 21–35). Woodhead Publishing. https://doi.org/10.1533/9781845692520.1.21
46. Fries, G. F. (1991). Organic Contaminants in Terrestrial Food Chains. In K. C. Jones (Ed.), *Organic Contaminants in the Environment: Environmental Pathways & Effects* (pp. 207–236). Dordrecht: Springer Netherlands. https://doi.org/10.1007/978-94-009-4329-2_7
47. Garcia-Garcia, G., Woolley, E., Rahimifard, S., Colwill, J., White, R., & Needham, L. (2017). A Methodology for Sustainable Management of Food Waste. *Waste and Biomass Valorization, 8*(6), 2209–2227. https://doi.org/10.1007/s12649-016-9720-0
48. Garcia-Herrero, I., Hoehn, D., Margallo, M., Laso, J., Bala, A., Batlle-Bayer, L., … Aldaco, R. (2018). On the estimation of potential food waste reduction to support sustainable production and consumption policies. *Food Policy, 80*, 24–38. https://doi.org/10.1016/j.foodpol.2018.08.007
49. García-Lomillo, J., González-SanJosé, M. L., Del Pino-García, R., Rivero-Pérez, M. D., & Muñiz-Rodríguez, P. (2014). Antioxidant and Antimicrobial Properties of Wine Byproducts and Their Potential Uses

in the Food Industry. *Journal of Agricultural and Food Chemistry, 62*(52), 12595–12602. https://doi.org/10.1021/jf5042678

50. Geng, Y., Zhu, Q., & Haight, M. (2007). Planning for integrated solid waste management at the industrial Park level: A case of Tianjin, China. *Waste Management, 27*(1), 141–150. https://doi.org/10.1016/j.wasman.2006.07.013
51. Ghose, B. (2014). Food security and food self-sufficiency in China: From past to 2050. *Food and Energy Security, 3*(2), 86–95. https://doi.org/10.1002/fes3.48
52. Girotto, F., Alibardi, L., & Cossu, R. (2015). Food waste generation and industrial uses: A review. *Waste Management, 45*, 32–41. https://doi.org/10.1016/j.wasman.2015.06.008
53. Giwa, A., Alabi, A., Yusuf, A., & Olukan, T. (2017). A comprehensive review on biomass and solar energy for sustainable energy generation in Nigeria. *Renewable and Sustainable Energy Reviews, 69*, 620–641. https://doi.org/10.1016/j.rser.2016.11.160
54. Gnansounou, E., & Dauriat, A. (2010). Techno-economic analysis of lignocellulosic ethanol: A review. *Bioresource Technology, 101*(13), 4980–4991. https://doi.org/10.1016/j.biortech.2010.02.009
55. Gómez-Estaca, J., Calvo, M. M., Álvarez-Acero, I., Montero, P., & Gómez-Guillén, M. C. (2017). Characterization and storage stability of astaxanthin esters, fatty acid profile and α-tocopherol of lipid extract from shrimp (*L. vannamei*) waste with potential applications as food ingredient. *Food Chemistry, 216*, 37–44. https://doi.org/10.1016/j.foodchem.2016.08.016
56. Hamidian, A. H., Ozumchelouei, E. J., Feizi, F., Wu, C., Zhang, Y., & Yang, M. (2021). A review on the characteristics of microplastics in wastewater treatment plants: A source for toxic chemicals. *Journal of Cleaner Production, 295*, 126480. https://doi.org/10.1016/j.jclepro.2021.126480
57. Han, F. X., Su, Y., Monts, D. L., Plodinec, M. J., Banin, A., & Triplett, G. E. (2003). Assessment of global industrial-age anthropogenic arsenic contamination. *Naturwissenschaften, 90*(9), 395–401. https://doi.org/10.1007/s00114-003-0451-2
58. Henry, R. K., Yongsheng, Z., & Jun, D. (2006). Municipal solid waste management challenges in developing countries – Kenyan case study. *Waste Management, 26*(1), 92–100. https://doi.org/10.1016/j.wasman.2005.03.007

59. Hjelm, L., Mielby, L. A., Gregersen, S., Eggers, N., & Bertram, H. C. (2019). Partial substitution of fat with rye bran fibre in Frankfurter sausages – Bridging technological and sensory attributes through inclusion of collagenous protein. *LWT*, *101*, 607–617. https://doi.org/10.1016/j.lwt.2018.11.055
60. Höök, M., & Tang, X. (2013). Depletion of fossil fuels and anthropogenic climate change—A review. *Energy Policy*, *52*, 797–809. https://doi.org/10.1016/j.enpol.2012.10.046
61. Howard, P. H. (2015). Intellectual Property and Consolidation in the Seed Industry. *Crop Science*, *55*(6), 2489–2495. https://doi.org/10.2135/cropsci2014.09.0669
62. Huang, C.-C., & Ma, H.-W. (2004). A multidimensional environmental evaluation of packaging materials. *Science of The Total Environment*, *324*(1), 161–172. https://doi.org/10.1016/j.scitotenv.2003.10.039
63. Hungate, R. E. (1982). Methane formation and cellulose digestion—Biochemical ecology and microbiology of the rumen ecosystem. In H. Mislin & R. Bachofen (Eds.), *New Trends in Research and Utilization of Solar Energy through Biological Systems* (pp. 117–120). Basel: Birkhäuser. https://doi.org/10.1007/978-3-0348-6305-6_22
64. Ishangulyyev, R., Kim, S., & Lee, S. H. (2019). Understanding Food Loss and Waste—Why Are We Losing and Wasting Food? *Foods*, *8*(8), 297. https://doi.org/10.3390/foods8080297
65. Islam, M. M., Sunny, A. R., Hossain, M. M., & Friess, D. A. (2018). Drivers of mangrove ecosystem service change in the Sundarbans of Bangladesh. *Singapore Journal of Tropical Geography*, *39*(2), 244–265. https://doi.org/10.1111/sjtg.12241
66. Jaggard, K. W., Qi, A., & Ober, E. S. (2010). Possible changes to arable crop yields by 2050. *Philosophical Transactions of the Royal Society B: Biological Sciences*, *365*(1554), 2835–2851. https://doi.org/10.1098/rstb.2010.0153
67. Jayathilakan, K., Sultana, K., Radhakrishna, K., & Bawa, A. S. (2012). Utilization of byproducts and waste materials from meat, poultry and fish processing industries: A review. *Journal of Food Science and Technology*, *49*(3), 278–293. https://doi.org/10.1007/s13197-011-0290-7
68. Jianyi, L., Yuanchao, H., Shenghui, C., Jiefeng, K., & Lilai, X. (2015). Carbon footprints of food production in China (1979–2009).

Journal of Cleaner Production, *90*, 97–103. https://doi.org/10.1016/j.jclepro.2014.11.072

69. Jingura, R. M., & Matengaifa, R. (2009). Optimization of biogas production by anaerobic digestion for sustainable energy development in Zimbabwe. *Renewable and Sustainable Energy Reviews*, *13*(5), 1116–1120. https://doi.org/10.1016/j.rser.2007.06.015
70. Jones, J. W., Antle, J. M., Basso, B., Boote, K. J., Conant, R. T., Foster, I., … Wheeler, T. R. (2017). Brief history of agricultural systems modeling. *Agricultural Systems*, *155*, 240–254. https://doi.org/10.1016/j.agsy.2016.05.014
71. Jouhara, H., Czajczyńska, D., Ghazal, H., Krzyżyńska, R., Anguilano, L., Reynolds, A. J., & Spencer, N. (2017). Municipal waste management systems for domestic use. *Energy*, *139*, 485–506. https://doi.org/10.1016/j.energy.2017.07.162
72. Kapoor, R., Ghosh, P., Kumar, M., Sengupta, S., Gupta, A., Kumar, S. S., … Pant, D. (2020). Valorization of agricultural waste for biogas based circular economy in India: A research outlook. *Bioresource Technology*, *304*, 123036. https://doi.org/10.1016/j.biortech.2020.123036
73. Kim, D., Kim, G., Oh, D. Y., Seong, K.-W., & Park, K. Y. (2022). Enhanced hydrogen production from anaerobically digested sludge using microwave assisted pyrolysis. *Fuel*, *314*, 123091. https://doi.org/10.1016/j.fuel.2021.123091
74. Kim, J. S., Lee, Y. Y., & Kim, T. H. (2016). A review on alkaline pretreatment technology for bioconversion of lignocellulosic biomass. *Bioresource Technology*, *199*, 42–48. https://doi.org/10.1016/j.biortech.2015.08.085
75. Kist, L. T., Moutaqi, S. E., & Machado, Ê. L. (2009). Cleaner production in the management of water use at a poultry slaughterhouse of Vale do Taquari, Brazil: A case study. *Journal of Cleaner Production*, *17*(13), 1200–1205. https://doi.org/10.1016/j.jclepro.2009.04.006
76. Klooster, D. (2010). Standardizing sustainable development? The Forest Stewardship Council's plantation policy review process as neoliberal environmental governance. *Geoforum*, *41*(1), 117–129. https://doi.org/10.1016/j.geoforum.2009.02.006
77. Kluts, I. N., Potting, J., Bosma, R. H., Phong, L. T., & Udo, H. M. J. (2012). Environmental comparison of intensive and integrated agriculture–aquaculture systems for striped catfish production in the

Mekong Delta, Vietnam, based on two existing case studies using life cycle assessment. *Reviews in Aquaculture*, *4*(4), 195–208. https://doi.org/10.1111/j.1753-5131.2012.01072.x

78. Kosseva, M. R. (2009). Chapter 3 Processing of Food Wastes. In *Advances in Food and Nutrition Research*: *Vol.58*. *Advances in Food and Nutrition Research* (pp. 57–136). Academic Press. https://doi.org/10.1016/S1043-4526(09)58003-5
79. Kumar, A., & Sharma, M. P. (2014). GHG emission and carbon sequestration potential from MSW of Indian metro cities. *Urban Climate*, *8*, 30–41. https://doi.org/10.1016/j.uclim.2014.03.002
80. Kummu, M., de Moel, H., Porkka, M., Siebert, S., Varis, O., & Ward, P. J. (2012). Lost food, wasted resources: Global food supply chain losses and their impacts on freshwater, cropland, and fertiliser use. *Science of The Total Environment*, *438*, 477–489. https://doi.org/10.1016/j.scitotenv.2012.08.092
81. Lal, R., Bouma, J., Brevik, E., Dawson, L., Field, D. J., Glaser, B., … Zhang, J. (2021). Soils and sustainable development goals of the United Nations: An International Union of Soil Sciences perspective. *Geoderma Regional*, *25*, e00398. https://doi.org/10.1016/j.geodrs.2021.e00398
82. Leal Filho, W., & Kovaleva, M. (2015a). Causes of Food Waste Generation. In W. Leal Filho & M. Kovaleva (Eds.), *Food Waste and Sustainable Food Waste Management in the Baltic Sea Region* (pp. 31–50). Cham: Springer International Publishing. https://doi.org/10.1007/978-3-319-10906-0_3
83. Leal Filho, W., & Kovaleva, M. (2015b). *Food Waste and Sustainable Food Waste Management in the Baltic Sea Region*. Cham: Springer International Publishing. https://doi.org/10.1007/978-3-319-10906-0
84. Lemos, P., & Mesquita, F. C. (2016). Chapter 11 - Future of Global Bioethanol: An Appraisal of Results, Risk and Uncertainties. In S. L. M. Salles-Filho, L. A. B. Cortez, J. M. F. J. da Silveira, S. C. Trindade, & M. da G. D. Fonseca (Eds.), *Global Bioethanol* (pp. 221–237). Academic Press. https://doi.org/10.1016/B978-0-12-803141-4.00011-3
85. Lin, C. S. K., Koutinas, A. A., Stamatelatou, K., Mubofu, E. B., Matharu, A. S., Kopsahelis, N., … Luque, R. (2014). Current and future trends in food waste valorization for the production of chemicals, materials and fuels: A global perspective. *Biofuels, Bioproducts and Biorefining*, *8*(5), 686–715. https://doi.org/10.1002/bbb.1506

86. Liu, C., Hotta, Y., Santo, A., Hengesbaugh, M., Watabe, A., Totoki, Y., … Bengtsson, M. (2016). Food waste in Japan: Trends, current practices and key challenges. *Journal of Cleaner Production, 133*, 557–564. https://doi.org/10.1016/j.jclepro.2016.06.026

87. Liu, X., Gao, X., Wang, W., Zheng, L., Zhou, Y., & Sun, Y. (2012). Pilot-scale anaerobic co-digestion of municipal biomass waste: Focusing on biogas production and GHG reduction. *Renewable Energy*, *44*, 463–468. https://doi.org/10.1016/j.renene.2012.01.092

88. Lusk, J. L., Roosen, J., & Fox, J. A. (2003). Demand for Beef from Cattle Administered Growth Hormones or Fed Genetically Modified Corn: A Comparison of Consumers in France, Germany, the United Kingdom, and the United States. *American Journal of Agricultural Economics*, *85*(1), 16–29. https://doi.org/10.1111/1467-8276.00100

89. Madilo, F. K., Owusu-Kwarteng, J., Parry-Hanson Kunadu, A., & Tano-Debrah, K. (2020). Self-reported use and understanding of food label information among tertiary education students in Ghana. *Food Control*, *108*, 106841. https://doi.org/10.1016/j.foodcont.2019.106841

90. Malak-Rawlikowska, A., Majewski, E., Wąs, A., Borgen, S. O., Csillag, P., Donati, M., … Wavresky, P. (2019). Measuring the Economic, Environmental, and Social Sustainability of Short Food Supply Chains. *Sustainability*, *11*(15), 4004. https://doi.org/10.3390/su11154004

91. Masarin, F., Gurpilhares, D. B., Baffa, D. C., Barbosa, M. H., Carvalho, W., Ferraz, A., & Milagres, A. M. (2011). Chemical composition and enzymatic digestibility of sugarcane clones selected for varied lignin content. *Biotechnology for Biofuels*, *4*(1), 55. https://doi.org/10.1186/1754-6834-4-55

92. McDonald, S., Prenzler, P. D., Antolovich, M., & Robards, K. (2001). Phenolic content and antioxidant activity of olive extracts. *Food Chemistry*, *73*(1), 73–84. https://doi.org/10.1016/S0308-8146(00)00288-0

93. McGlade, C., & Ekins, P. (2015). The geographical distribution of fossil fuels unused when limiting global warming to 2 °C. *Nature*, *517*(7533), 187–190. https://doi.org/10.1038/nature14016

94. McKenzie, F. C., & Williams, J. (2015). Sustainable food production: Constraints, challenges and choices by 2050. *Food Security*, *7*(2), 221–233. https://doi.org/10.1007/s12571-015-0441-1

95. Melikoglu, M., Lin, C. S. K., & Webb, C. (2013a). Analysing global food waste problem: Pinpointing the facts and estimating the energy content. *Central European Journal of Engineering*, *3*(2), 157–164. https://doi.org/10.2478/s13531-012-0058-5
96. Melikoglu, M., Lin, C. S. K., & Webb, C. (2013b). Kinetic studies on the multi-enzyme solution produced via solid state fermentation of waste bread by Aspergillus awamori. *Biochemical Engineering Journal*, *80*, 76–82. https://doi.org/10.1016/j.bej.2013.09.016
97. Meynard, J.-M., Dedieu, B., & Bos, A. P. (Bram). (2012). Re-design and co-design of farming systems. An overview of methods and practices. In I. Darnhofer, D. Gibbon, & B. Dedieu (Eds.), *Farming Systems Research into the 21st Century: The New Dynamic* (pp. 405–429). Dordrecht: Springer Netherlands. https://doi.org/10.1007/978-94-007-4503-2_18
98. M'hiri, N., Veys-Renaux, D., Rocca, E., Ioannou, I., Boudhrioua, N. M., & Ghoul, M. (2016). Corrosion inhibition of carbon steel in acidic medium by orange peel extract and its main antioxidant compounds. *Corrosion Science*, *102*, 55–62. https://doi.org/10.1016/j.corsci.2015.09.017
99. Mirabella, N., Castellani, V., & Sala, S. (2014). Current options for the valorization of food manufacturing waste: A review. *Journal of Cleaner Production*, *65*, 28–41. https://doi.org/10.1016/j.jclepro.2013.10.051
100. Mittal, G. S. (2006). Treatment of wastewater from abattoirs before land application—A review. *Bioresource Technology*, *97*(9), 1119–1135. https://doi.org/10.1016/j.biortech.2004.11.021
101. Mohee, R., & Mudhoo, A. (2005). Analysis of the physical properties of an in-vessel composting matrix. *Powder Technology*, *155*(1), 92–99. https://doi.org/10.1016/j.powtec.2005.05.051
102. Möllersten, K., Yan, J., & R. Moreira, J. (2003). Potential market niches for biomass energy with CO2 capture and storage—Opportunities for energy supply with negative CO2 emissions. *Biomass and Bioenergy*, *25*(3), 273–285. https://doi.org/10.1016/S0961-9534(03)00013-8
103. Moruzzo, R., Riccioli, F., Boncinelli, F., Zhang, Z., Zhao, J., Tang, Y., … Guidi, A. (2020). Urban Consumer Trust and Food Certifications in China. *Foods*, *9*(9), 1153. https://doi.org/10.3390/foods9091153
104. Murdoch, J., & Miele, M. (1999). 'Back to Nature': Changing 'Worlds of Production' in the Food Sector. *Sociologia Ruralis*, *39*(4), 465–483. https://doi.org/10.1111/1467-9523.00119

105. Obi, F. O., Ugwuishiwu, B. O., & Nwakaire, J. N. (2016). Agricultural waste concept, generation, utilization and management. *Nigerian Journal of Technology*, *35*(4), 957–964. https://doi.org/10.4314/njt.v35i4

106. Olsson, M. E., Gustavsson, K.-E., & Vågen, I. M. (2010). Quercetin and Isorhamnetin in Sweet and Red Cultivars of Onion (*Allium cepa* L.) at Harvest, after Field Curing, Heat Treatment, and Storage. *Journal of Agricultural and Food Chemistry*, *58*(4), 2323–2330. https://doi.org/10.1021/jf9027014

107. Papamonioudis, K., & Zabaniotou, A. (2022). Exploring Greek Citizens' Circular Thinking on Food Waste Recycling in a Circular Economy—A Survey-Based Investigation. *Energies*, *15*(7), 2584. https://doi.org/10.3390/en15072584

108. Pappu, A., Saxena, M., & Asolekar, S. R. (2007). Solid wastes generation in India and their recycling potential in building materials. *Building and Environment*, *42*(6), 2311–2320. https://doi.org/10.1016/j.buildenv.2006.04.015

109. Pode, R. (2016). Potential applications of rice husk ash waste from rice husk biomass power plant. *Renewable and Sustainable Energy Reviews*, *53*, 1468–1485. https://doi.org/10.1016/j.rser.2015.09.051

110. Popp, J., Harangi-Rákos, M., Gabnai, Z., Balogh, P., Antal, G., & Bai, A. (2016). Biofuels and Their Co-Products as Livestock Feed: Global Economic and Environmental Implications. *Molecules*, *21*(3), 285. https://doi.org/10.3390/molecules21030285

111. Porwal, H. J., Mane, A. V., & Velhal, S. G. (2015). Biodegradation of dairy effluent by using microbial isolates obtained from activated sludge. *Water Resources and Industry*, *9*, 1–15. https://doi.org/10.1016/j.wri.2014.11.002

112. Pour, N., Webley, P. A., & Cook, P. J. (2017). A Sustainability Framework for Bioenergy with Carbon Capture and Storage (BECCS) Technologies. *Energy Procedia*, *114*, 6044–6056. https://doi.org/10.1016/j.egypro.2017.03.1741

113. Pun, K. F., Chin, K. S., & Lau, H. (1999). A self-assessed quality management system based on integration of MBNQA/ISO 9000/ISO 14000. *International Journal of Quality & Reliability Management*, *16*(6), 606–629. https://doi.org/10.1108/02656719910268206

114. Rahmanian, N., Jafari, S. M., & Galanakis, C. M. (2014). Recovery and Removal of Phenolic Compounds from Olive Mill Wastewater. *Journal of the American Oil Chemists' Society*, *91*(1), 1–18. https://doi.org/10.1007/s11746-013-2350-9
115. Rao, P. V., Baral, S. S., Dey, R., & Mutnuri, S. (2010). Biogas generation potential by anaerobic digestion for sustainable energy development in India. *Renewable and Sustainable Energy Reviews*, *14*(7), 2086–2094. https://doi.org/10.1016/j.rser.2010.03.031
116. Ravi Kumar, K., Krishna Chaitanya, N. V. V., & Sendhil Kumar, N. (2021). Solar thermal energy technologies and its applications for process heating and power generation – A review. *Journal of Cleaner Production*, *282*, 125296. https://doi.org/10.1016/j.jclepro.2020.125296
117. Reilly, J. M. (2012). Green growth and the efficient use of natural resources. *Energy Economics*, *34*, S85–S93. https://doi.org/10.1016/j.eneco.2012.08.033
118. Rogelj, J., Luderer, G., Pietzcker, R. C., Kriegler, E., Schaeffer, M., Krey, V., & Riahi, K. (2015). Energy system transformations for limiting end-of-century warming to below 1.5 °C. *Nature Climate Change*, *5*(6), 519–527. https://doi.org/10.1038/nclimate2572
119. Roupas, P., De Silva, K., Smithers, G., & Ferguson, A. (2007). 13—Waste management and co-product recovery in red and white meat processing. In K. Waldron (Ed.), *Handbook of Waste Management and Co-Product Recovery in Food Processing* (pp. 305–331). Woodhead Publishing. https://doi.org/10.1533/9781845692520.4.305
120. Roy, H., Alam, S. R., Bin-Masud, R., Prantika, T. R., Pervez, M. N., Islam, M. S., & Naddeo, V. (2022). A Review on Characteristics, Techniques, and Waste-to-Energy Aspects of Municipal Solid Waste Management: Bangladesh Perspective. *Sustainability*, *14*(16), 10265. https://doi.org/10.3390/su141610265
121. Ryder, K., Ali, M. A., Billakanti, J., & Carne, A. (2020). Evaluation of Dairy Co-product Containing Composite Solutions for the Formation of Bioplastic Films. *Journal of Polymers and the Environment*, *28*(2), 725–736. https://doi.org/10.1007/s10924-019-01635-4
122. Sabir, A., Altaf, F., Batool, R., Shafiq, M., Khan, R. U., & Jacob, K. I. (2021). Agricultural Waste Absorbents for Heavy Metal Removal. In Inamuddin, M. I. Ahamed, E. Lichtfouse, & A. M. Asiri (Eds.), *Green Adsorbents to Remove Metals, Dyes and Boron from Polluted Water*

(pp. 195–228). Cham: Springer International Publishing. https://doi.org/10.1007/978-3-030-47400-3_8

123. Sadiq, M., Mayyas, A., Mezher, T., & El Fadel, M. (2023). Policy and economic challenges towards scalable green-H2 transition in the middle east and north Africa region. *International Journal of Hydrogen Energy*. https://doi.org/10.1016/j.ijhydene.2023.05.083

124. Schrader, K. K., de Regt, M. Q., Tidwell, P. D., Tucker, C. S., & Duke, S. O. (1998). Compounds with selective toxicity towards the off-flavor metabolite-producing cyanobacterium *Oscillatoria* cf. Chalybea. *Aquaculture*, *163*(1), 85–99. https://doi.org/10.1016/S0044-8486(98)00223-3

125. Sharma, K., & Garg, V. K. (2018). Chapter 17 - Solid-State Fermentation for Vermicomposting: A Step Toward Sustainable and Healthy Soil. In A. Pandey, C. Larroche, & C. R. Soccol (Eds.), *Current Developments in Biotechnology and Bioengineering* (pp. 373–413). Elsevier. https://doi.org/10.1016/B978-0-444-63990-5.00017-7

126. Sheldon, R. A., Bode, M. L., & Akakios, S. G. (2022). Metrics of green chemistry: Waste minimization. *Current Opinion in Green and Sustainable Chemistry*, *33*, 100569. https://doi.org/10.1016/j.cogsc.2021.100569

127. Smil, V. (2016). Examining energy transitions: A dozen insights based on performance. *Energy Research & Social Science*, *22*, 194–197. https://doi.org/10.1016/j.erss.2016.08.017

128. Stalikas, C. D., & Konidari, C. N. (2001). Analytical methods to determine phosphonic and amino acid group-containing pesticides. *Journal of Chromatography A*, *907*(1), 1–19. https://doi.org/10.1016/S0021-9673(00)01009-8

129. Sun, Y., Tsao, R., Chen, F., Li, H., Wang, J., Peng, H., … Deng, Z. (2017). The phytochemical composition, metabolites, bioavailability and *in vivo* antioxidant activity of *Tetrastigma hemsleyanum* leaves in rats. *Journal of Functional Foods*, *30*, 179–193. https://doi.org/10.1016/j.jff.2017.01.004

130. Swain, P. S., Dominic, G., Bhakthavatsalam, K. V. S., & Terhuja, M. (2016). Impact of Ruminants on Global Warming: Indian and Global Context. In S. Nautiyal, R. Schaldach, K. V. Raju, H. Kaechele, B. Pritchard, & K. S. Rao (Eds.), *Climate Change Challenge (3C) and Social-Economic-Ecological Interface-Building: Exploring Potential Adaptation Strategies for Bio-resource Conservation and Livelihood*

Development (pp. 83–97). Cham: Springer International Publishing. https://doi.org/10.1007/978-3-319-31014-5_6

131. Tekin, A. R., & Dalgıç, A. C. (2000). Biogas production from olive pomace. *Resources, Conservation and Recycling*, *30*(4), 301–313. https://doi.org/10.1016/S0921-3449(00)00067-7
132. Teo, M. M. M., & Loosemore, M. (2001). A theory of waste behaviour in the construction industry. *Construction Management and Economics*, *19*(7), 741–751. https://doi.org/10.1080/01446190110067037
133. Thapa, P., Hasnine, M. T., Zoungrana, A., Thakur, S., & Yuan, Q. (2022). Food Waste Treatments and the Impact of Composting on Carbon Footprint in Canada. *Fermentation*, *8*(10), 566. https://doi.org/10.3390/fermentation8100566
134. Tripathi, N., Hills, C. D., Singh, R. S., & Atkinson, C. J. (2019). Biomass waste utilisation in low-carbon products: Harnessing a major potential resource. *Npj Climate and Atmospheric Science*, *2*(1), 1–10. https://doi.org/10.1038/s41612-019-0093-5
135. Turan, N. G., Çoruh, S., Akdemir, A., & Ergun, O. N. (2009). Municipal solid waste management strategies in Turkey. *Waste Management*, *29*(1), 465–469. https://doi.org/10.1016/j.wasman.2008.06.004
136. van der Ploeg, J. D., Jingzhong, Y., & Schneider, S. (2012). Rural development through the construction of new, nested, markets: Comparative perspectives from China, Brazil and the European Union. *The Journal of Peasant Studies*, *39*(1), 133–173. https://doi.org/10.1080/03066150.2011.652619
137. van Dijk, M., Morley, T., Rau, M. L., & Saghai, Y. (2021). A meta-analysis of projected global food demand and population at risk of hunger for the period 2010–2050. *Nature Food*, *2*(7), 494–501. https://doi.org/10.1038/s43016-021-00322-9
138. Vijayan, D. S., & Parthiban, D. (2020). Effect of Solid waste based stabilizing material for strengthening of Expansive soil- A review. *Environmental Technology & Innovation*, *20*, 101108. https://doi.org/10.1016/j.eti.2020.101108
139. Voll, G. K., Manfred. (1993). Manufacture of Carbon Black. In *Carbon Black* (2nd ed.). Routledge.
140. Wang, F. C.-Y. (2000). Polymer additive analysis by pyrolysis–gas chromatography: IV. Antioxidants. *Journal of Chromatography A*, *891*(2), 325–336. https://doi.org/10.1016/S0021-9673(00)00647-6

141. Wang, L., Weller, C. L., Jones, D. D., & Hanna, M. A. (2008). Contemporary issues in thermal gasification of biomass and its application to electricity and fuel production. *Biomass and Bioenergy*, *32*(7), 573–581. https://doi.org/10.1016/j.biombioe.2007.12.007
142. Wang, T., Li, Q., & Bi, K. (2018). Bioactive flavonoids in medicinal plants: Structure, activity and biological fate. *Asian Journal of Pharmaceutical Sciences*, *13*(1), 12–23. https://doi.org/10.1016/j.ajps.2017.08.004
143. Wheeler, S. (2013). *Planning for Sustainability: Creating Livable, Equitable and Ecological Communities* (2nd ed.). London: Routledge. https://doi.org/10.4324/9780203134559
144. Wunderlich, S. M., & Martinez, N. M. (2018). Conserving natural resources through food loss reduction: Production and consumption stages of the food supply chain. *International Soil and Water Conservation Research*, *6*(4), 331–339. https://doi.org/10.1016/j.iswcr.2018.06.002
145. Yang, Q., Yang, M., Zhang, S., & Lv, W. (2005). Treatment of wastewater from a monosodium glutamate manufacturing plant using successive yeast and activated sludge systems. *Process Biochemistry*, *40*(7), 2483–2488. https://doi.org/10.1016/j.procbio.2004.09.009
146. Yeung, R. M. W., & Morris, J. (2006). An empirical study of the impact of consumer perceived risk on purchase likelihood: A modelling approach. *International Journal of Consumer Studies*, *30*(3), 294–305. https://doi.org/10.1111/j.1470-6431.2006.00493.x
147. Yokoo, H.-F., Kawai, K., & Higuchi, Y. (2018). Informal recycling and social preferences: Evidence from household survey data in Vietnam. *Resource and Energy Economics*, *54*, 109–124. https://doi.org/10.1016/j.reseneeco.2018.08.001
148. Zhang, N., Bai, Z., Luo, J., Ledgard, S., Wu, Z., & Ma, L. (2017). Nutrient losses and greenhouse gas emissions from dairy production in China: Lessons learned from historical changes and regional differences. *Science of The Total Environment*, *598*, 1095–1105. https://doi.org/10.1016/j.scitotenv.2017.04.165
149. zu Ermgassen, E. K. H. J., Phalan, B., Green, R. E., & Balmford, A. (2016). Reducing the land use of EU pork production: Where there's swill, there's a way. *Food Policy*, *58*, 35–48. https://doi.org/10.1016/j.foodpol.2015.11.001

INDEX

B

C

D

E

I

J

L

M

Q

R

S

T

U

V